IREX III

Performance of Iris Identification Algorithms

NIST Interagency Report 7836

P. Grother   G. W. Quinn   J. R. Matey

M. Ngan   W. Salamon   G. Fiumara   C. Watson

Information Access Division

National Institute of Standards and Technology

April 12, 2012

Acknowledgements

The authors would like to thank the sponsors of this activity. These are the Federal Bureau of Investigation, the U.S. VISIT office in the Department of Homeland Security (DHS), and the Science and Technology Directorate, also in DHS.

Vitally, the authors would also like to thank the United States Department of Defense's Biometrics Identity Management Agency for their support and active collaboration.

Finally, the authors are grateful to Dr. Mark Burge (MITRE) for his constructive review of this document.

Disclaimer

Specific hardware and software products identified in this report were used in order to perform the evaluations described in this document. In no case does identification of any commercial product, trade name, or vendor, imply recommendation or endorsement by the National Institute of Standards and Technology, nor does it imply that the products and equipment identified are necessarily the best available for the purpose.

Contents

List of Figures

FNIR = FALSE NEGATIVE IDENT. RATE	N = NEUROTECHNOLOGY	P = SMU	Q = IRITECH	R = COGENT	S = SMARTSENSORS	T = CAMBRIDGE
FPIR = FALSE POSITIVE IDENT. RATE	U = L1	V = MORPHO	W = IRISID	X = CROSSMATCH	Y = KYNEN	

List of Tables

Executive Summary

The Iris Exchange IREX III evaluation was conducted to measure the accuracy and speed of iris identification algorithms. The test was designed to have operational relevance. First, by using as many as 6.1 million images of 4.3 million eyes, the results are relevant to systems used for the full range one-to-many applications including de-duplication, benefits fraud, and token-less access and border control. Second the test ran algorithms on commodity PC-class blade computers running the LINUX operating system which are typical in central-server applications. Third, the test applied algorithms to archival *operational* data, thereby mimicing a real-world task. The algorithms were invoked to do one-to-many searches in a database of enrolled iris images to produce lists of candidate identities sorted in increasing order of dissimilarity value. Two kinds of searches are executed. The first, searches with an enrolled mate, allow measurement and reporting of the core false negative identification rate, i.e. the "miss" rate. The second, searches for which there is no enrolled mate, support measurement of false positive identification rate, more commonly known as "false alarm" rate. These quantities are estimated as a function of enrolled population sizes.

While the test is primarily useful for comparative evaluation of algorithms, it is also pertinent to algorithm developers and those concerned with operational deployment of iris identification.

This report is accompanied by the IREX III FAILURE ANALYSIS supplement which documents the causes of recognition "misses" and suggests methods to avert these. Also of note is the November 2011 IREX II IQCE[30] report on the capability of dedicated iris image quality algorithms to measure image properties that are related both miss and false alarm outcomes. All IREX documents are linked from http://iris.nist.gov/irex.

The main conclusions of the IREX III evaluation are:

▷ **Absolute iris recognition accuracy**: Over the 95 algorithms evaluated, single-eye iris identification false negative error rates (i.e. "miss" rates) are at 1.5% or higher. For two eyes, this figure is 0.7%. These figures constitute an accuracy floor that reflects the presence of hard and, in some cases, impossible to "hit" images. The failures comprising this floor arise from abnormal irides, from poor quality images of normal irides (e.g. those that have non-axial gaze, pupil constriction, large rotation, eyelid occlusions, blur, environmental reflections, and camera artifacts), from defective preparation and storage of images (quantization and compression artifacts), and from ground-truth errors (multiple persons sharing one ID). These exigencies are documented here and in the IREX III FAILURE ANALYSIS supplement that accompanies this report and is intended to assist implementers in mitigation and remediation of random and systematic iris collection and recognition errors. Note IREX III was conducted with capable, yet legacy, cameras. Current and future iris recognition deployments, which almost universally use two-eye cameras, can reduce failures by appropriate design and selection of cameras, environments, collection procedures, clerical controls, and user instruction.   Section 6.3

▷ **Low false positive rates**: As with all biometric algorithms, false positive rates can be made arbitrarily low by applying stringent decision thresholds. However, the tradeoff between false negative and false positive error rates, as reported on the *detection error tradeoff* (DET) characteristics, means that false negative errors become untenably large for weak biometric modalities, poor samples, and inaccurate algorithms. Iris recognition has long had a reputation for producing very low false positive rates. For the most accurate algorithm evaluated here recognizing single irises in a population of size 3.9 million, the use of a threshold set to produce no more than 25 false matches in every 10^{13} comparisons yields a false negative identification error rate below 2.5%. Other specific false positive rates can be targeted by adopting thresholds given in calibration curves reported for each algorithm. These are computed from up to 1.2 trillion unmated comparisons, offering the best published means to target particular (low) false positive rates.   Section 6.9

▷ **Comparative algorithm accuracy**: On a large fixed set of test images, false negative ("miss") rates span an order of magnitude depending on the implementation being evaluated. Thus in the scenario of the previous paragraph, while the most accurate algorithm produces a false negative rate below 2.5%, the miss rate for the less accurate algorithms can greatly exceed 20%. Moreover, some iris algorithms evaluated here have DETs with low gradient; this allows iris to be used in large populations, or equivalently with low false positive rates, while suffering only small increases in false negatives.   Section 6.5

FNIR = FALSE NEGATIVE IDENT. RATE	N = NEUROTECHNOLOGY	P = SMU	Q = IRITECH	R = COGENT	S = SMARTSENSORS	T = CAMBRIDGE
FPIR = FALSE POSITIVE IDENT. RATE	U = L1	V = MORPHO	W = IRISID	X = CROSSMATCH	Y = KYNEN	

▷ **Iris recognition in investigations**: The same algorithms can be used in an investigational mode with relaxed thresholds. Here they fail to return the correct mate at rank 1 in only about 1.5% of searches, and at rank 20 in about 1.2% of searches. However, because false positives are present on the candidate lists, an investigative application assumes and requires trained human reviewers to adjudicate images. While this process might begin with manual segmentation or image enhancement and re-execution of the algorithms, it moves on to visual comparison of iris textural features (such as crypts) and then devolves to inspection of such as eye lash roots, eye brows and the shapes of the two canthi (if visible). The stability and, particularly, the uniqueness of these structures, however, has not been measured. Indeed, because the forensic discipline of human review of iris imagery extends to little more than the text of this paragraph, the efficacy of combined iris-human systems is uncertain. Note that the human adjudication of iris pairs can be difficult even when image quality is good. This occurred in IREX III when the most similar of 1.2 trillion nonmate pairs were inspected.   Section 7.3

▷ **Algorithm development**: A major factor in the success of future deployments will be the improvement and selection of iris recognition algorithms. The wide accuracy variations observed here demonstrates the need for algorithm providers to continue research and development. Particularly, we document algorithm-idiosyncratic failure modes that should be addressed by specific developers. These fall into two categories, those that affect both *false negative* performance, for example the handling of dilated eyes, but also *false positive* effects, for example the handling of overly compressed images. Improvements may become evident in the upcoming IREX IV activity.   Section 8.1

▷ **Feature representations**: Templates embed proprietary feature data extracted from iris images. While IREX III regarded templates as black box data, it did measure their size and generation time. The two most accurate implementations produce enrollment templates of size 1 and 20 kilobytes (KB) respectively. Enrollment templates varied in size from 257 bytes to 20008 bytes, with most occupying less than 10KB. These sizes differ from those published for the original Cambridge University *iriscode* feature set indicating that there is abundant additional and independent diversity in iris representation. Indeed the 1KB Cambridge template size measured in this study is larger than that reported in the seminal Cambridge publications.

While small templates support one-to-many searches on mobile devices and fast network-based search, no standard iris recognition template has ever been formalized by ISO committees, NIST, or any other recognized standards organization. Template-based deployments are not future-proofed against technical developments, and are subject to a non-interoperable vendor lock-in hazard. These problems can be averted by use and/or retention of standard *images* (ISO/IEC 19794-6:2011) which, given specialized preparation, can occupy as little as 2-3KB for 1:1, and 50-80KB for 1:N.

Three providers' algorithms generate search templates (which typically exist only for the duration of a recognition transaction) that are larger than the enrollment templates.   Section 4.1

▷ **Scalability**: When a threshold is fixed for an identification algorithm, the widely held model of biometric identification accuracy states that the False Positive Identification Rate (FPIR) grows linearly with population size while the False Negative Identification Rate (FNIR) remains unchanged. However, while this is found to be correct for many of the algorithms evaluated here, some algorithms exhibit false positive rates that are essentially *independent* of population size at a fixed operating threshold. This result, which applies to algorithms from two IREX III participants, is unknown in the biometrics literature. This valuable property, which comes without elevated false negative errors and without measureable increase in computational cost, relieves system operators of the need to adjust the operating threshold as subjects are enrolled. This capability is consistent with introduction of an explicit dependence on N, or the use of (score) normalization techniques. Either would be implemented within the identification search engine, and may be configurable - one vendor submitted implementations for which FPIR is dependent on N and others where it is independent. The availability and efficacy of this feature from other providers is not known. For other algorithms and larger population sizes, the stability of FPIR should be measured, not assumed.   Section 6.8.2

▷ **Two-eye operation**: For most algorithms operated with a fixed threshold, two-eye searches against a population of two-eye enrollees give approximately half as many false negative errors (i.e. misses) compared to single-eye operation. However, because recognition failure of a left eye comparison is correlated with that of a right eye comparison (e.g. both eyes are occluded by long eyelashes or both eyes are looking left), the benefit from use of two eyes is less than that implied from naïve independence statistics. Fingerprint recognition is similarly affected. The reduced FNIR is accompanied by approximately four times as many false positives compared to the case when a single eye is used. The elevated false positive rates occur because of the internal 2x2 cross comparison of two search eyes with two enrolled eyes. This implies that the algorithms are implementing fusion mechanisms equivalent to OR-decision or MIN-score fusion. That is to say, a match is returned if either of the search irides match either of the enrolled irides. This reporting of the lowest dissimilarity score has been considered in the academic literature[27].

Note, however, that this observed result holds when left-right eye labels are unknown to the algorithm. In operational cases where eye label information is available and reliable, only a two-fold increase in false positive identification rate is expected. However there is a further caveat here: In an operation where the prior probability of a mate is low (i.e. not the IREX III test), false match performance might be improved by fusing according to AND- or MAX-score rules. This would rely on the published observation that a person's left-and-right eye textures are as dissimilar as those from unrelated persons, such that both would have to falsely match to declare a false positive.

Importantly, some algorithms combine reduced false negatives without any increase in false positive rates for two-eye recognition. Section 6.6

▷ **Image quality values**: Even before completion of a standardized definition of iris image quality (ISO/IEC 29794-6, now under development), low quality values from commercial implementations are being used to contraindicate enrollment and matching. In IREX III some template generation algorithms produce a scalar image quality value that is some proprietary combination of measurements such as such as exposed iris area, blur, pupil constriction and dilation, and gaze deviation. The November 2011 IREX II - IMAGE QUALITY CALIBRATION AND EVALUATION report considered this in great detail. In IREX III the utility of a quality value is assessed in terms of its relation to recognition outcome. The result is that the most capable quality assessor assigns low quality values to only 23.6% of image pairs involved in the poorest 2% of false negative outcomes (i.e. highest dis-similarity scores). For other implementations this figure is as low as 2.5%. Together these figures mean that rejection based on low image quality values would not avert 76.4% and 97.5% of these identification misses. Note that the IREX III template generators do not necessarily embed the dedicated image quality assessment algorithms considered in IREX II, and so dedicated quality assessment algorithms might improve this situation. Section 8.5

▷ **Pupil constriction and dilation**: Iris images with constricted and dilated pupils give elevated recognition error rates. Both false negative and false positive errors are related to the ratio of pupil and iris radii. In decreasing order of severity, some mated searches give more false negatives when: the pupils are differently dilated, are both constricted, or are both dilated. False negative identification rates for mated pairs increase exponentially in the change in radial thickness of the iris texture.

In nonmate searches, most algorithms have an adverse dependence on dilation. That is, nonmate dissimilarity scores tend to be lower when either the pupil in the search image or the enrolled image is dilated. Only a few algorithms give the behavior for constricted pupils. While these behaviors have been reported for smaller datasets in IREX I and, for single algorithms, in the academic literature, dilation has been more problematic than constriction. The importance of constricted pupils in IREX III is due to the systematically more constricted pupils present versus that reported for laboratory data. Binocular cameras, intended to mitigate pupil constriction by shielding the eyes from ambient light, are effective at giving a narrower range of pupil dilation and reduced error rates. Section 8.3

▷ **Algorithm efficiency**: Iris identification speeds span at least three orders of magnitude. This massive variation reflects an industry-wide diversity of algorithms and parameterizations of those algorithms. Two factors drive

FNIR = FALSE NEGATIVE IDENT. RATE	N = NEUROTECHNOLOGY	P = SMU	Q = IRITECH	R = COGENT	S = SMARTSENSORS	T = CAMBRIDGE
FPIR = FALSE POSITIVE IDENT. RATE	U = L1	V = MORPHO	W = IRISID	X = CROSSMATCH	Y = KYNEN	

this. First, the IREX III API explicitly called for demonstration of speed-accuracy tradeoffs, and second, the public nature of the NIST tests drives developers to push the limits of both accuracy *and* speed. In particular, any given provider is likely to have submitted their mature mainline operational algorithms and also their experimental prototypes. The result, for the two most accurate implementations, is that the most computationally expensive algorithm affords no accuracy benefit over one that is about 400 times faster. These two algorithms do, however, produce about one third fewer false negative errors than several other algorithms that run five or six times faster.

Template generation times vary by a factor of forty, rising from 0.02 to nearly 0.8 seconds[1]. The faster algorithms can easily run on mobile and wall mounted camera hardware. This is important in allowing on-camera recognition. For some algorithms, template generation represents the largest contribution to the total time needed to take an image, prepare the template and execute the 1:N search. While the 1:N search duration eventually dominates the total time, the breakeven population, at which search time exceeds template generation time, is nearly four million for some algorithms.

On the same PC-hardware, the fastest algorithm executes around eight million matches per second per core. This algorithm gives double the number of false negative identification errors as the most accurate algorithms which execute only several hundred thousand comparisons per second. An important qualifier here is that absolute speed is generally subject to a very large number of factors associated with hardware, architecture, processor instructions, compilers, threading, optimization, algorithm choice, enrolled population size, and heuristic short cuts. Rather the purpose here is to expose algorithmic complexity and relative computational cost, and to support comparative evaluation.   Section 5

▷ **Comparison with face recognition**: Using an almost identical protocol to that used in IREX III, NIST evaluated one-to-many face recognition algorithms in the MBE 2010 program[11]. Both studies used images from operational detainee registration processes: face images from persons arrested in law enforcement context, and iris images from individuals encountered during military interdiction. Given this reasonable real-world basis for modality comparison, single-iris identification gives an order of magnitude fewer errors than that for single-face search: In an enrolled population of 1.6 million, with thresholds set to produce a false positive in every 16 billion comparisons iris gives a factor of ten fewer misses than face (2% vs. 20%). Two-iris operation would double this improvement. The shape of the respective DETs indicates that iris will give at least 100,000 times fewer false positives than face, for an equal false negative identification rate. However, when thresholds are relaxed and forensic face and iris examiners would be relied upon to adjudicate candidate lists, iris offers about a factor of four fewer misses. In summary, iris recognition is viable for automated identification in parts of the accuracy tradespace that are not accessible using single-image face recognition.

The leading iris algorithms are faster on average than face, but there are large speed variations, and the most accurate face algorithm is faster than one of the most accurate iris algorithms. Also, some face algorithms show better-than-linear dependence of speed on enrolled population size.   Appendix A.3

While accuracy and speed are necessary to the technical success of most biometric applications, a large number of other considerations will drive biometric modality selection. These include speed and ease of biometric capture, societal and economic factors, the existence of legacy databases, policy drivers, marketplace size and maturity, and availability of, support for, and conformance to, standards.   Appendix A.4

▷ **Confusion of retina and iris recognition**. A number of sources mistake retinal biometrics for iris. The iris and retina are anatomically distinct structures at the front and back of the eye respectively, and are imaged using different optical designs.

---

[1] All timing estimates in this report are measured on standard PC-class hardware holding 64-bit AMD processors manufactured c. 2008. The precise hardware and software specifications were fixed and fully disclosed prior to the test in the IREX III API, CONOPS AND EVALUATION PLAN.

| FNIR = FALSE NEGATIVE IDENT. RATE | N = NEUROTECHNOLOGY | P = SMU | Q = IRITECH | R = COGENT | S = SMARTSENSORS | T = CAMBRIDGE |
| FPIR = FALSE POSITIVE IDENT. RATE | U = L1 | V = MORPHO | W = IRISID | X = CROSSMATCH | Y = KYNEN | |

Release Notes

All IREX related reports, drafts, announcements and news items may be found on the homepage
http://iris.nist.gov/irex.

▷ **Concept of Operations**: IREX III was conducted in accordance with the IREX III API, CONOPS, AND EVALUATION SPECIFICATION. This was developed by NIST in consultation with members of the iris recognition community. The document was drafted in October 2010, circulated for public comment on two occasions, and finalized on February 11, 2011. The API was adapted from one used for a large-scale face recognition test conducted in 2010[11].

▷ **Supplemental report**: This report is also being released with a companion IREX III SUPPLEMENTAL that is intended to document non-ideal images and best practices recommendations for their avoidance.

▷ **Appendices**: This report is accompanied by the IREX III APPENDICES, which present exhaustive results on a per-algorithm basis. The document, which extends to several hundred pages, is machine generated. It will be of primary interest to the algorithm developers, and to users considering particular algorithms.

▷ **Algorithm identifiers**: Throughout this report the implementations are identified by alphanumeric code of the form xGDU with uppercase letter $x \in \{N \ldots Y\}$ identifying the provider of the SDK, G is a single digit identifying a group, D being a sequence number that distinguishes submissions from the same provider, and U being the *claimed* class of participation. Class A was intended to denote a fast implementation, and B a slower or more experimental entry. In the end the labels were vendor-defined and are not particularly meaningful. Group 0 indicates an algorithm was submitted to IREX III from February to June 2011, and Group 1 in August 2011. The codes support automated administration of the test, and conserve space in the tables of this report. **For reference, the letters are associated with the providers' names in a running footnote.**

▷ **Typesetting**: Virtually all of the tabulated content in this report was produced automatically. This involved the use of scripting tools to generate directly typesettable LaTeX content. This improves timeliness, flexibility, maintainability, and reduces transcription errors.

▷ **Contact**: Correspondence regarding this report should be directed to PGROTHER at NIST dot GOV.

FNIR = FALSE NEGATIVE IDENT. RATE	N = NEUROTECHNOLOGY	P = SMU	Q = IRITECH	R = COGENT	S = SMARTSENSORS	T = CAMBRIDGE
FPIR = FALSE POSITIVE IDENT. RATE	U = L1	V = MORPHO	W = IRISID	X = CROSSMATCH	Y = KYNEN	

The Iris Exchange (IREX) Program

In 2008 NIST established the IREX program to give quantitative support to iris recognition standardization, development and deployment. The activities that have been conducted under IREX so far are:

▷ **IREX I**: The 2009 IREX I evaluation, which tested the efficacy of leading commercial and university algorithms on the specialized image formats proposed for the ISO/IEC 19794-6 iris image data interchange standard. IREX I also established viable limits for standardized image compression algorithms applied to iris images. Accuracy was measured over one-to-one comparisons.

▷ **IREX II**: The 2010-2011 Iris Quality Calibration and Evaluation (IQCE), which assessed the capabilities of iris image quality assessment algorithms and supported the ISO/IEC 29794-6 iris image quality standard by establishing metrics, reference thresholds, and ranges for various appearance, geometric and photometric properties of iris images. Accuracy was measured using one-to-one comparisons operating separately from the image quality assessment algorithm.

▷ **IREX III**: The 2011 IREX III activity, which is the analysis of one-to-many iris performance documented in this report.

▷ **IREX IV**: The 2012 IREX IV activity, proposed as an direct follow on to the IREX III study, will apply contemporary one-to-many recognition algorithms to newly available uncompressed iris images. This work will support development of definitive JPEG 2000 compression profiles for iris identification. This extends the IREX I work by considering the false positive demands of one-to-many, and by refining JPEG 2000's parameters.

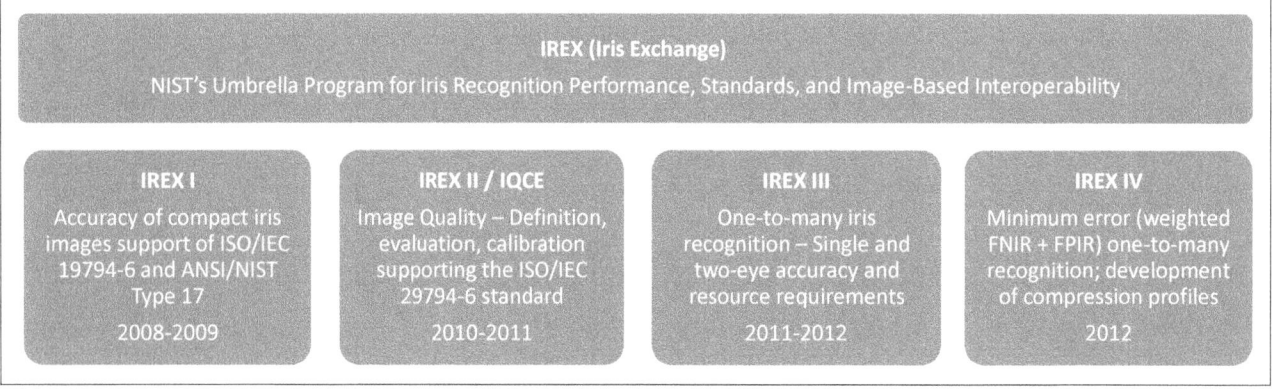

Figure 1: The four phases of the IREX program.

Caveats

Biometric test results have limited relevance outside of the context in which they are obtained, and therefore caveats apply to the quantitative results and conclusions. The main limitations preventing IREX III from being universally relevant to one-to-many identification performance are as follows.

1. **Specific nature of the iris data**: The absolute error rates quoted here were measured over a very large fixed corpus of operational iris images. The error rates measured here are realistic if the algorithms were applied to this kind of data. However, in other applications, the applicability of the results may differ due to a number of factors legitimately not reflected in the IREX III experimental design. Among these are:

   ▷ Capture environment - Outdoor implementations, especially in hot environments with direct sunlight, will fare poorly due to infrared contamination of the scene. Some cameras (e.g. binocular-style) shield the sensors from ambient lighting, while others do not. The IREX III imagery was collected in a variety of physical environments, some not completely shielded from external light sources - see the IREX III FAILURE ANALYSIS supplement.

   ▷ Subject cooperation - If the subject is not incentivized, or is actively dis-incentivized, to use the camera in the intended mode, then failures-to-capture and/or false rejections will increase. The IREX III images were collected from detainees who, for the most part, were neither motivated to cooperate nor evade iris collection. That said, the collection protocol required the capture of an iris.

   ▷ Camera improvements - Late model cameras likely have improved imaging capabilities over the cameras presented here. This might arise from better designs of optics, image quality assessment software, and the increased computational capability of processors embedded in cameras. Much of the IREX III imagery was collected with L1 cameras that internally require larger-than-normal areas of the iris to be exposed. This necessitated using fingers to hold eyes open, for which the Indian Government has reported accuracy benefits[2].

   ▷ Camera interoperability - Use of a single camera, by definition, removes possible cross-camera interoperability issues. These can occur because different cameras produce different images from the same iris. Differences may arise from variations in the wavelength and incidence angles of the IR illumination, and potentially other imaging properties of the device. IREX III data was collected with multiple camera models from several providers.

   ▷ Single- or two-eye imaging - Use of single-eye cameras can give rise to left-right eye label inversion errors and rotation of the camera about the optical axis. While most contemporary cameras image both eyes and avoid these problems, some that implement truly simultaneously capture of left and right eyes may give higher incidence of problems such as blinks, non-frontal gaze angle, and motion blur. Most IREX III data was collected with single-eye cameras. The exception is the subset whose dimensions are 480x480 pixels.

2. **Research and development**: Any test of algorithms is subject to a shelf-life limitation on the relevance of the results (the methods persist longer). In particular, given large-scale operational deployments in India, Indonesia and elsewhere, the accuracy and related cost imperatives are instrumental in driving development.

3. **Version control**: Some algorithms were specialized experimental variants intended solely to demonstrate good performance on the IREX III test. These may be unavailable or inappropriate for general-purpose applications. Moreover, the ability of some providers to recall and re-implement any particular algorithm may be imperfect.

4. **Platform**: Computational limitations may prevent some of IREX III algorithms from being ported and used on less powerful hardware. For example 64-bit addressing is not universal, but is needed for large population identification.

5. **Non-iris constraints**: Some applications will have operational aspects that might limit attainable performance. For example, if images must be heavily compressed to satisfy a communications-channel limitation[3], then accuracy may degrade. Similarly if only one eye was imaged, then accuracy may degrade.

---

[2]See http://uidai.gov.in/images/FrontPageUpdates/uid_enrolment_poc_report.pdf
[3]For which dedicated standardized formats exist in ISO/IEC 19794-6:2011.

FNIR = FALSE NEGATIVE IDENT. RATE	N = NEUROTECHNOLOGY	P = SMU	Q = IRITECH	R = COGENT	S = SMARTSENSORS	T = CAMBRIDGE
FPIR = FALSE POSITIVE IDENT. RATE	U = L1	V = MORPHO	W = IRISID	X = CROSSMATCH	Y = KYNEN	

1   Introduction

Biometric identification systems account for the largest fraction of revenue from sales of recognition technologies. This reflects their broad applicability and the non-trivial nature of the problem. The core one-to-many search function supports identification-mode applications ranging from straightforward criminal search[4], through watchlist detection and reverse search[5], benefits fraud de-duplication[6], social network tagging[7], and access control[8]. Due to an increase in the frequency at which false positives occur, identification becomes more difficult as the enrolled population size grows. Thus, while gymnasiums often use a single finger to allow access for its several thousand members, the government of India's Unique Identity (UID) program is using ten-fingers and two-irides to assign de-duplicated numerical identifiers to hundreds of millions of residents[9]. The diversity of identification applications is responsible for one-to-many functionality being the largest segment of the biometric recognition marketplace.

Iris recognition has long been held as a powerful biometric suited to accurate identification[4]. It has recently been supported by expanded availability of advanced cameras that are more portable, capture irides much more quickly, and do so at substantially greater distances[24]. However, there is a paucity of experimental data to support published theoretical considerations and accuracy claims. Particularly, prior studies have either not been independent[21, 25], have considered only one-to-one verification[12], used small populations[17], or have not been public. The IREX III study documented here addresses this gap by applying algorithms developed in commercial and academic research laboratories to a dataset of several million irides, and quantifying accuracy and speed. The study constitutes the first public presentation of results for iris identification algorithms tested independently using an operational dataset. The study additionally compares one-to-many face and iris identification performance. It does not compare iris with fingerprints.

2   Algorithm submission and use

Participation in IREX III was open to any commercial, academic, or non-profit organization as well as individuals. The algorithm providers are listed in Table 1. The only necessary qualifications were those implied by the requirement to implement the interface given in the IREX III API, CONOPS, AND EVALUATION PLAN. This necessitated only possession

---

[4]Systems allowing search of a crime-scene latent fingerprint against a convict repository are used worldwide by most policing agencies.

[5]For example, the United Arab Emirates border crossing authorities use iris recognition to detect immigration fraud, including prior deportees.

[6]Most United States' driving license issuers use face recognition to detect drivers whose license has been previously revoked and who apply under a pseudonym.

[7]Various social networking sites suggest candidate names for friends whose faces appear in uploaded images.

[8]For example, in the CANPASS system at several Canadian airports and the IRIS system at London Heathrow, border crossing is expedited by positively identifying a traveller's irides against the database of previously registered travellers. This is done in one-to-many mode, without presentation of a passport or other token.

[9]http://uidai.gov.in/images/FrontPageUpdates/role_of_biometric_technology_in_aadhaar_jan21_2012.pdf

Participant Name	Letter Code	Num SDKs Reported
Cambridge University	T	10
Cogent Systems	R	8
Crossmatch Technologies	X	5
IrisID	W	10
Iritech	Q	8
Kynen	U	4
L1 Identity Solutions	U	10
Morpho	V	9
Neurotechnology	N	8
Smartsensors	S	8
Southern Methodist University (SMU)	P	2

Table 1:   IREX III providers. The number of implementations reported in some cases differs from the actual number tested because some early implementations were inoperable or slow. The maximum number of implementations allowed was 10. Not all providers elected to submit that many.

---

of iris recognition algorithms and software engineering skills sufficient to implement specific C++ API calls and data structures.

Algorithms were submitted to NIST as static (".a") or dynamic link (".so") libraries compiled for execution on a recent LINUX kernel. The sizes of these files are presented in Appendix J in the companion IREX III APPENDICES document. Arbitrary files containing configuration or training data optionally accompanied the implementation.

Some implementations used a common template representation. That is, all templates from the following algorithms were identical. These are:

▷ N02A, N02B, N03A, N04A, N11A, N12A

▷ N12B, N13B

▷ Q03B, Q04B

▷ Q02A, Q02B

▷ R03A, R11A, R12A

▷ S11A, S12A

▷ T03A, T03B, T04A

▷ T11A, T11B, T12B

▷ T01B, T02B

▷ U11B, U12A, U12B

▷ U03A, U04A

▷ U01A, U02A

▷ U01B, U02B

▷ W11A, W12A

▷ W02A, W04A

In the case of U01A and U02A, the templates are almost identical, but the candidate lists from searches are identical, per section 4.3. The accuracy and search time results for the above algorithms are due solely to search, comparison and indexing algorithms. The template generation durations given later in Table 8 were computed for all algorithms independently and show variations in these cases within statistical confidence limits.

	Nominal Pop	Num People	Num Enrolled Identities	Num Images
Single Eye	20000	10173	20001	20001
	160000	81415	160000	160000
	1600000	819350	1600000	1600000
	4000000	1999989	3904239	3904239
Two Eye	20000	20000	20000	39312
	160000	160000	160000	313332
	1600000	1600000	1600000	3123677
	1	2	3	4

Table 2: Enrolled population sizes. This report refers to the enrolled populations identified in column two. These are target values drawn at random from the entire parent corpus. **For single-eye operation**, the number of human subjects in column three is approximately half of that number because left and right eyes are used as though they are independent biometric samples. The peak number of enrolled identities (3904239) is short of the four million target because a left or right eye image was not available from all persons. **For two-eye operation**, the number of images is slightly less than twice the population size because a left or right eye image was not available from all persons. In all cases the yellow shading indicates that the column is the value most relevant to biometric false match potential.

3   Preparation and use of the test data

The test was conducted with a single large iris image corpus consisting of $6,142,289$ images of $4,333,745$ eyes from $2,212,342$ human subjects. Of these, about 80% of subjects were imaged on only one occasion, the remainder being imaged multiple times, with about 1% being imaged ten or more times. Approximately 5% of captures resulted in images of only one eye of a subject.

The subjects were divided into two sets.

▷ The first set, $\mathcal{I}$, consists of $212,342$ subjects, and was reserved for conducting nonmate searches. The number of images in this set is $413,254$.

	Set Name	Num People	Num Searches	Num Mated Searches	Num Images
Single Eye	S_{1a}	15088	36748	17017	36748
	S_{1b}	231408	553230	238740	553230
Two Eye	S_{2a}	20000	27276	17276	53432
	S_{2b}	250361	315662	155662	617473
	1	2	3	4	5

Table 3: Numbers of searches conducted. This report refers to the search sets identified in column one. The number of searches is reported in column three, with the number of mated searches in column four. The number of mateless searches is the subtraction of column four from three. **For single-eye operation**, the number of human subjects in column two is less than half of the number of searches because some subjects were imaged on multiple (different) days. The peak number of enrolled identities (3904239) is short of the four million target because a left or right eye image was not available from some persons. **For two-eye operation**, images of a person are combined in the MULITIRIS datastructure of the IREX III API - this requires the SDK to implement search and fusion strategies. The number of images is slightly less than twice the population size because a left or right eye image was not available from some persons. In all cases the yellow shading indicates that the column is the value most relevant to high confidence estimation of false negative and postive identification error rates.

> ▷ The second set, $\mathcal{E}$, consists of the $1,999,989$ subjects, and was reserved for construction of enrollment sets and corresponding mated-search sets. The number of images in this set is $3,904,239$. Eleven subjects from the parent population were omitted because the images or ground-truth information were unreadable or unavailable.

3.1 Enrollment and search sets

The tables and figures of this report refer to population sizes $N \leq 3,904,239$. These correspond to the single- and two-eye enrollment sets $\mathcal{E}_N$ defined as follows.

> ▷ **Single-eye** The enrolled population sizes, $20,000 \leq N \leq 3,904,239$ refer to the enrollment of N single-eyes, typically left and right eyes from approximately $N/2$ people - The exact numbers are tabulated in Table 2. The enrollment sets are formed by randomly sampling subjects from the parent population $\mathcal{E}$. The number of images is not exactly double the number of subjects, because left and right eye images are not always available from some subjects. Note some figures refer to a population of 4 million; this figure is nominal, the actual size is 3,904,239 because of the unavailability of two eye images from some persons.

> ▷ **Two-eye** The enrolled population sizes, $20,000 \leq N \leq 1,600,000$ refer to the enrollment of N persons for whom either both left and right eyes are available or (rarely) for whom only a sole left or right is present. (This design avoids the possible bias introduced by only selecting images from persons for whom only one image is present.) The subjects are drawn randomly from $\mathcal{E}$. The iris images of each subject are passed to the SDK together in the MULTIIRIS data structure of the IREX III API. This allows the SDK to combine data from two eyes in any manner it sees fit. Typically this would involve separate feature extraction from each eye, but could involve some implementer-defined template-level fusion.

In all cases, only some enrollees have a subsequent mate. Also, if the enrolled set from population N is $\mathcal{E}_N$, the smaller sets are strict subsets of the larger $\mathcal{E}_n \in \mathcal{E}_N \quad \forall n < N$.

One-to-many search proceeds using the four search sets defined in Table 3. For both single- and two-eye cases, small and large sets are defined. The small sets (sizes in the tens of thousands) afford a *quick-look* estimation of performance. The larger sets give greater fidelity to those estimates and allow for covariate analysis.

3.2 Sets for estimation of function durations

The first 1000 images of set S_{1a} were used for estimating single-eye search speed. This subset contains images from 488 mate searches and 512 nonmate searches. Durations were measured for population sizes from 20,000 to 3.9 million, as

Camera	S	N_{P}	N_{PE}	$1 - N_{\mathrm{PE}}/N_{\mathrm{P}}$
BIDS	24547	24483	23794	0.03
COGT	176	146	128	0.12
L1-B	90792	88215	72715	0.18
L1-I	70688	69458	14583	0.79
LSMS-330	8221	7116	4031	0.43
LSMS-640	32439	32020	29930	0.07
MOBS-A	8077	7985	7928	0.01
MOBS-D	2057	1987	1776	0.11

Table 4: For each camera system, the number of searches S, the number of searches for which algorithm V11B hit the correct person, N_P, the number where it hit the correct person *and* eye, N_{PE}, and in the final column, the implied proportion of incorrect eye labels. This value is very similar for other recognition algorithms. This experimental design included two images for most individuals, typically one left and one right. The algorithms enrolled and identified images without eye label information. Two elements are highlighted to indicate discussion in the text. Some rows for other infrequently used cameras have been omitted.

detailed in Table 2. Note that because there is some dependence on the imagery itself, timing estimates would be improved slightly if the measurements were made over many randomly chosen search sets *and* enrollment sets.

3.3   Use of eye labels

Left and right eye labels for all images accompanied the images of all parent corpora. However, while the IREX III API supported provision of the eye labels to the template generators, this facility was not normally used. Thus, unless specified otherwise, all images were labeled "unknown". This was done for two reasons.

▷ **Unreliable LR labels**: First, some 30% of the images were assigned the incorrect left-right eye label[10]. This high proportion is a result of the use of single-eye cameras and the ease with which a human-operator, prompted to collect the left eye image first, actually collects the right. This confusion arises because the subject's left eye is on the camera operator's right. Single eye iris cameras do not attempt to automatically determine left-and-right. Table 4 shows the approximate prevalence of eye mislabelling for the cameras included in the OPS dataset. The green-shaded item corresponds to a two-eye camera. The reason the value is not identically zero is not known. The red-shaded item corresponds to a single-eye camera. The high value indicates a systematic problem in the eye-label metadata.

▷ **Inconsistent with test goals**: Second, while eye labels can be used to expedite search (e.g. given a left eye, search against left eyes only, realize a factor-of-two increase in speed), this is not essential to the measurement of speed of the underlying technology.

The consequences of not providing eye labels to the algorithms are material. First, without eye labels, a comparison of two images against an enrollment entry also containing two images will usually embed four template matches (L-L, L-R, R-L, R-R). This will immediately incur a four-fold increase in computational duration compared to single-eye operation, and a two-fold increase in duration compared to the two-eye case where eye labels are present and correct (L-L, R-R). The accuracy implications are discussed in section 6.6.

3.4   Use of camera identifier

The IREX III API enumerated integer camera identifiers for 18 different commercial iris cameras. Camera identifiers can be used by informed developers to constrain spatial and rotational extent in iris-detection algorithms, and possibly to augment other processing. In the current study however, camera identifiers were not provided to the template generator. Instead all images were assigned a "camera unknown" default. This denies those implementers who are aware of camera-specific variables the opportunity to improve performance. It also reduces the possible bias introduced for those IREX III participants who might also have manufactured the cameras used in the collection of the operational data.

---

[10]This figure was estimated using the more accurate algorithms as the fraction of mated searches finding the correct person minus the fraction identifying the correct person *and* eye (according to the ground truth). The first fraction is typically FNIR≤ 0.05; the second is 0.3 larger.

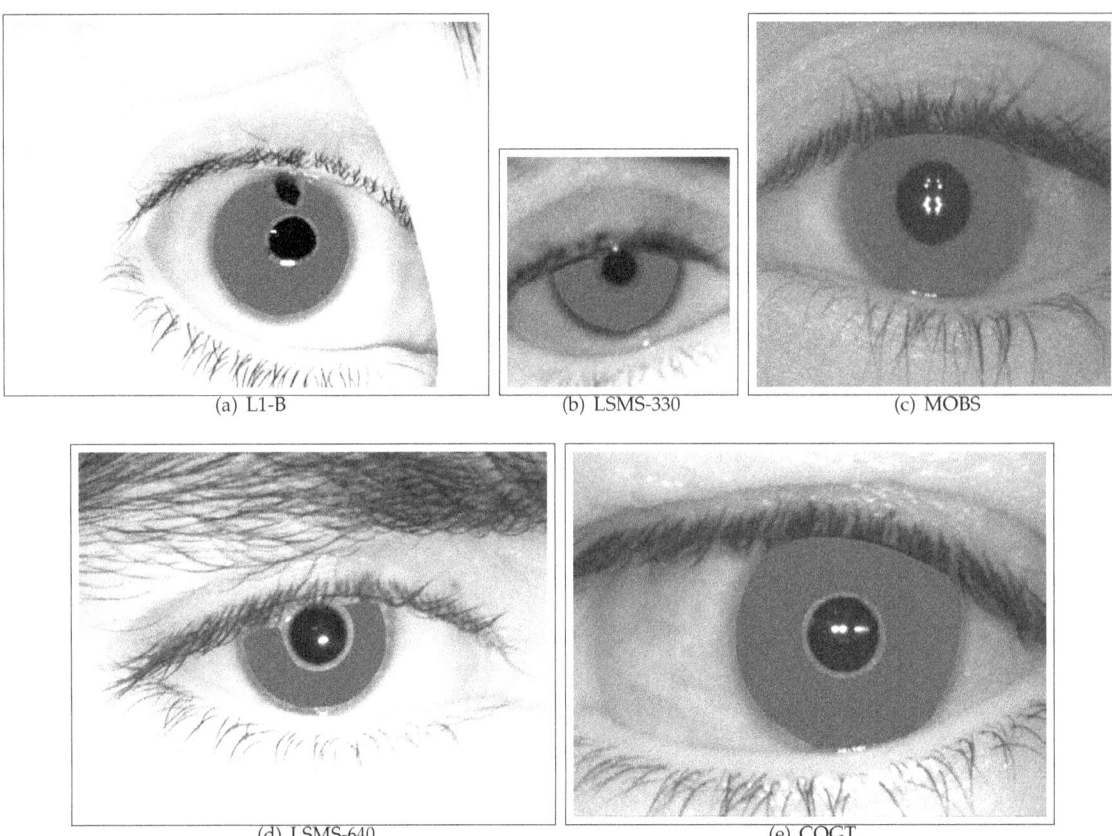

Figure 2: Selected images from various cameras. The red covering is applied by NIST to de-identify the iris. Note that any presentation of a few images cannot hope to capture quality-related variations that result from optical design tradeoffs such as illuminant power, integration time, head movement tolerance and threshold, depth-of-field range, compliance, etc. The shape of specular reflections from the LED illuminants is characteristic of the camera. The design is such that, given good user presentation, the reflections will appear on the pupil and be sharp - cameras often use the reflection for focus control. Holding eyes open, as in the first image, is common, and is intended to expedite capture and improve exposed iris area. This practice has been reported worthwhile in India's UID enrollment. While the fingers are not always visible in the scene, highly curved eyelids often are.

3.5   Provision of image type information

The IREX III API's image type ID was set to 0, connoting image size 640x480 pixels. The test proceeded with images with some images of size 480x480 and 330x330 pixels. This algorithms functioned without problem, but NIST nevertheless informed the algorithm providers that non-640x480 images were being used. There were no objections.

3.6   Degraded images

One subset of the operational data has images of size 330x330 pixels. The iris is well-centered in these images. In many cases, these images exhibit severe JPEG compression tiling artifacts. In addition, specular reflections from the camera illuminant have been masked to black. An example is shown in Figure 3(a). JPEG compression is not allowed in formal iris image standards, viz. ANSI/NIST ITL 1-2011 and ISO/IEC 19794-6:2011.

Other images, of size 640x480, exhibit evidence of grey level quantization. The process by which this occurs appears to have been intended to preserve iris texture while reducing the number of grey levels elsewhere to a small number. An example is shown in Figure 3(b). This method for supporting compression is non-standard, and should be strongly deprecated in favor of the compact formats established in ISO/IEC 19794-6:2011 and tested in IREX I.

| FNIR = FALSE NEGATIVE IDENT. RATE | N = NEUROTECHNOLOGY | P = SMU | Q = IRITECH | R = COGENT | S = SMARTSENSORS | T = CAMBRIDGE |
| FPIR = FALSE POSITIVE IDENT. RATE | U = L1 | V = MORPHO | W = IRISID | X = CROSSMATCH | Y = KYNEN | |

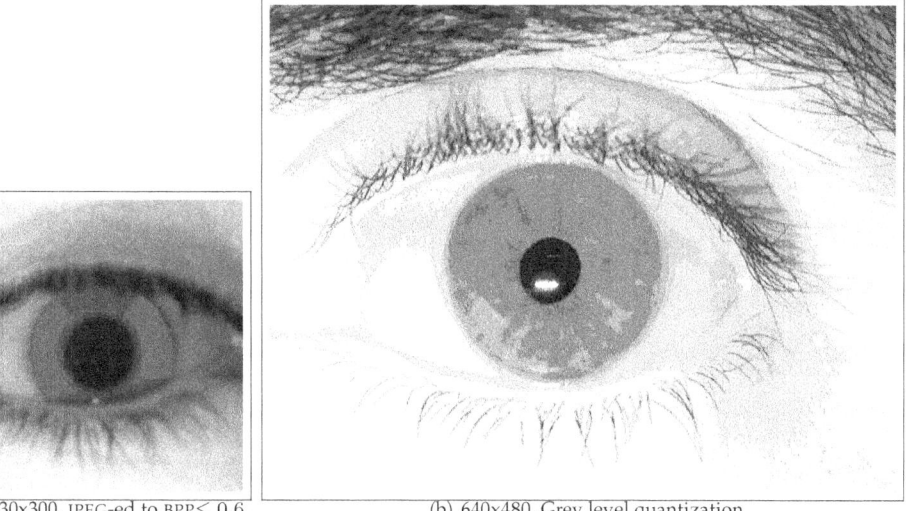

(a) 330x300, JPEG-ed to BPP≤ 0.6.   (b) 640x480, Grey level quantization

Figure 3:   Pathological images: Examples of overly compressed images (left) and quantized images (right). The red covering is applied by NIST to de-identify the iris. Subject to printing and display limitations the reader should see JPEG tiling artifacts in the left-hand image. The potential for such misuse of JPEG has motivated the prohibition of JPEG from formal standards. In both images the iris texture has been severely corrupted and, as such, the images are case studies for how not to prepare iris images for recognition. The 330x330 images were excluded from most analyses in this report.

3.7   Use of multiple cameras

The IREX III dataset includes images from a mixture of cameras. Much of the effort toward standardization of iris images has been to establish an interoperable format. This extends beyond mere definition of a syntax for containers[11] to a specification of images that can be produced by camera A and consumed by recognition algorithm B, where A and B are not manufactured by the same commercial entity. This aspect underpins the emergence of a plural marketplace of interoperating cameras.

False Negative identification Rates By Camera Pair									Count By Camera Pair		
U12A	CM	L1	N12A	CM	L1	RGA	CM	L1	Count	CM	L1
CM	0.015	0.018	CM	0.024	0.036	CM	0.046	0.053	CM	15011	26373
L1	0.015	0.034	L1	0.027	0.053	L1	0.042	0.059	L1	22292	108721

Table 5:   The table shows, for three algorithms, FNIR at FPIR = 0.0001 for intra- and inter-operability of two families of cameras CM (Crossmatch Technologies) and L1 (L1 Identity Solutions). In the rightmost two columns are the counts of the number of mated pairs. These results are potentially undermined, as a formal statement of interoperability, by lack of controls for: operational role, location, population, environment, upgrades and variation in hardware, firmware and software.

Table 5 shows interoperability of the main camera families used in IREX III. These are the SEEK devices from Crossmatch Technologies (CM), and the PIER and HIIDE cameras from L1 Identity Solutions (L1). The results here, and for all other algorithms in IREX III APPENDICES, show that the CM cameras produce lower error rates than those from L1 and that cross-camera accuracy falls between these two values. These observations are purely empirical, and may very well have to do with pupil dilation rather than anything to do with the optical and electronic design.

---

[11] ISO/IEC 19794-6:2011 defines a binary image format containing a general header, one or more image representation headers, and one or more images rasters encoded according to ISO/IEC 15444 and ISO/IEC 15948. Such records could come directly from a camera. ANSI/NIST ITL 1-2011 is equivalent.

FNIR = FALSE NEGATIVE IDENT. RATE	N = NEUROTECHNOLOGY	P = SMU	Q = IRITECH	R = COGENT	S = SMARTSENSORS	T = CAMBRIDGE
FPIR = FALSE POSITIVE IDENT. RATE	U = L1	V = MORPHO	W = IRISID	X = CROSSMATCH	Y = KYNEN	

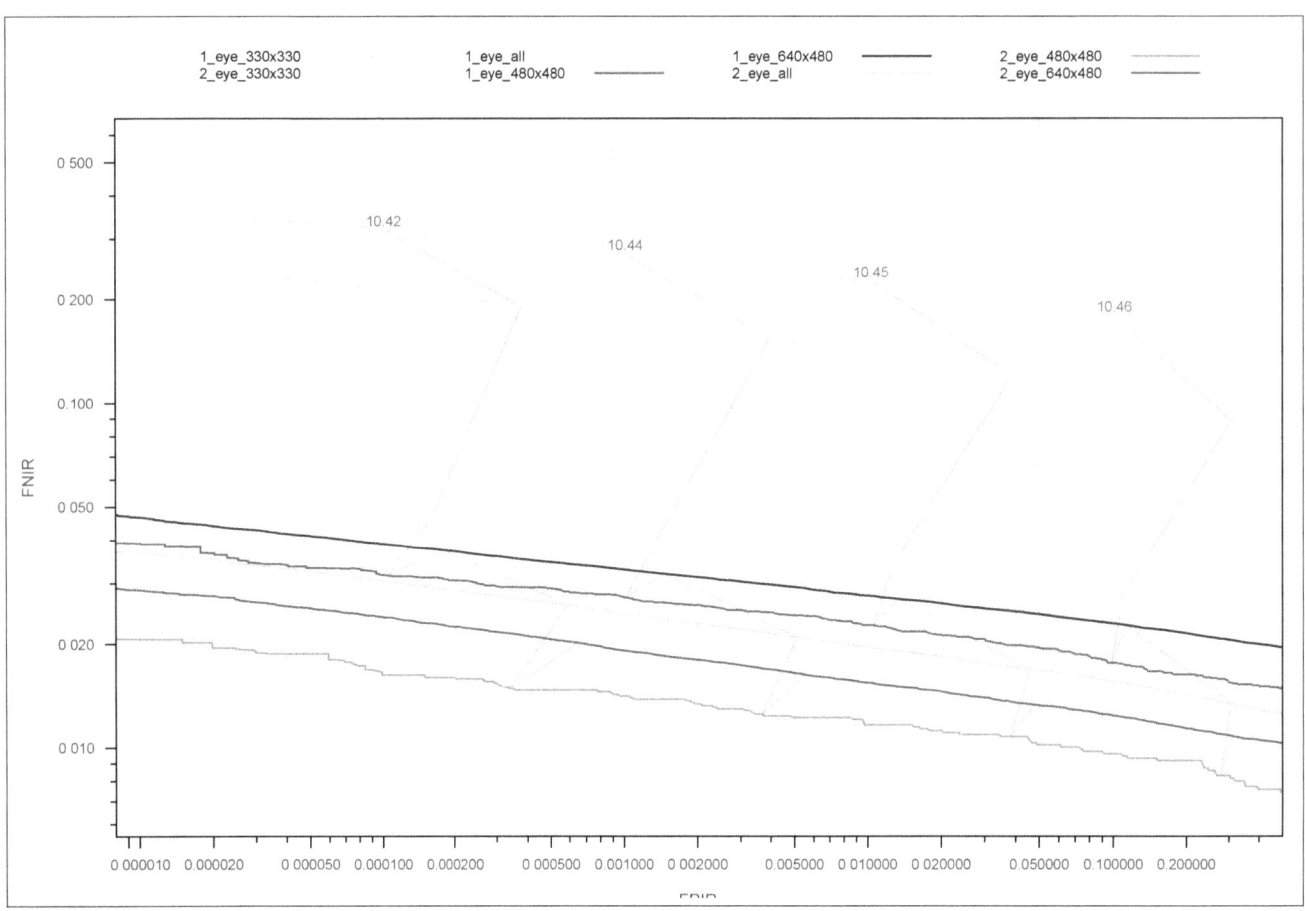

Figure 4: DET for the U12A implementation executing searches from the large search sets, S_{xb} in enrolled populations of N = 1.6M one- and two-eye identities. The eight plots show FNIR as a function of FPIR for all combinations of 330x300, 480x480, 640x480 pixel images, and all-sizes pixel images, for one and two eyes. The straight lines, which link points of fixed threshold, reveal variations in *both* FNIR and FPIR. Analogous figures for all implementations are presented in Appendix H (in the IREX III APPENDICES).

3.8 Use of cameras and algorithms from same providers

The IREX III test used images collected using several camera models (see section 3.4 and example images in Figure 2), collected using several camera models (see section 3.4), and some of these were manufactured by companies that are also IREX III participants. This raises the issue that the test has a bias toward certain algorithms.

The following points counter those concerns.

▷ **Other cameras work well too**: As shown in Figure 4, the L1 algorithms are more accurate on the 480x480 images (not L1 cameras) than on the 640x480 images (mostly from L1 cameras). See the algorithm-specific DETs in Appendix H (in the IREX III APPENDICES).

▷ **Separated subsidiaries**: Many of the images were collected using cameras manufactured in California by the Securimetrics subsidiary of L1. The PIER and HIIDE cameras may have included quality assurance procedures, for example image quality assessment or cross comparison of several collected images. The L1 recognition algorithms submitted to this test originated at L1's office in New Jersey. Whether the algorithms resident in the Securimetrics cameras pre-date or share common lineage with any algorithms submitted here, is unknown.

▷ **No prior knowledge**: The number and origins of the test corpus was not announced prior to commencement of the test. Indeed the IREX III API document enumerated camera codes for 18 cameras from 9 commercial providers. However, prospective participants, as watchers of the U.S. Government and the marketplace, might have assumed NIST would have access to U.S. Government imagery. That said, they might reasonably have thought NIST would have access to images from other sources.

▷ **Tuning risks**: Given that NIST did not disclose sources, the fine tuning of algorithms to specific cameras would present a possible hazard, because blind specialization might have hurt generality.

▷ **Access to data**: It is typical for governmental and other organizations that collect biometric imagery to not share it with their technology providers. The confidentiality of subjects, and images thereof, is usually protected from unauthorized access by policy, contracts, or law. The extent to which IREX participants had institutional exposure and actionable insight into the operational properties of the data is not known.

▷ **No camera identifiers**: As stated in section 3.4, IREX III algorithms were not provided with the camera identifier used for image collection. While camera information could possibly be recovered from watermarks (that survived image compression) or from the configuration of the LED reflections present in the image, this is unlikely.

4   Resource requirements

The section documents methods and results for the resource costs associated with the use of iris recognition algorithms. This includes measurements of template size and computational expense. To estimate the speed, the IREX III test harness wrapped all IREX III API functions with calls to the *gettimeofday* function. Under LINUX this timer has microsecond resolution. Given that the fastest template generators execute in tens of milliseconds, the worst case timing error is below 0.01%. For search, the duration of the call depends on N. For the smallest population size of $N = 20,000$, the shortest observed duration was $1100\mu s$, such that measurement error is well below 1%.

4.1   Template size

Each implementation encodes information derived from the iris image in a proprietary representation of the feature data. This information is generally a trade secret. It very likely encodes some mathematical representation of the iris texture[25, 6] but could also, in principle, encode anything else[12].

There is no formally standardized iris recognition template.

The size of the feature data is an important system-design parameter in most biometric applications. It has implications for permanent storage, in-memory storage, network transmission, and machine throughput.

Tables 6 and 7 show template sizes, in bytes. For two-eye inputs, the implementations produce a single binary template. This will usually embed features extracted from both eyes, but could include only data from one eye, or some combined, fused representation.

The headline observations are as follows.

> **Variation**: The smallest enrollment templates are 257 bytes (V04A) and 290 bytes (U03A, U04A). The largest enrollment templates exceed 20KB (V11B). The two most accurate algorithms use templates of size about 1KB (U12B) and 20KB (V12B). Six participating organizations submitted algorithms for which templates were about 1KB or less (N, S, T, U, V, W, and X).

> **Two-eyes**: Two-eye templates are universally almost exactly twice the size of single-eye templates. That is, implementers do nothing more than store both single-iris representations.

> **Variable-length templates**: Some implementations (e.g. S01B, N04A) produce templates whose length appears to vary. These templates likely include fixed-length and variable-length portions.

> **In-memory representation**: The in-memory organization of templates is implementation-defined; fixed length or lightweight feature data may be stored discontiguously and separately from the remainder. The *on-disk* storage values of Tables 6 and 7 are sometimes larger than that of the enrollment templates. This is because the IREX III API allowed implementations to *finalize* the enrollment by preparing arbitrary data structures for direct loading into memory ahead of search.

> **Asymmetric templates**: Search templates from some Sxxx algorithms and for W12A are an order of magnitude larger than the enrollment templates. For some Uxxx algorithms, the search templates are two or four times larger, and, for most Wxxx implementations, the factor is about two. Many operational systems only store enrollment templates, making the size of the search template only relevant if templates are being transmitted (instead of, or in addition to, images) from the collection point to a backend recognition system.

---

[12] The following have been identified as novel biometrics in the academic literature or in patent filings: eye lashes and roots, periocular skin texture, and the capillaries visible in the sclera. Their inclusion in IREX III templates would represent a high risk proposition for the implementer because, without dedicate acquisition devices, the availability of the signal was highly uncertain ahead of the test.

	Single Eye									Two eyes				
	Enrollment N = 160000					Search S = 553230				Enrollment N = 160000			Search S = 315662	
	Template				Disk	Template				Template		Disk	Template	
	Mode	Second	Mean	SD.	Mean	Mode	Second	Mean	SD.	Mode	Mean	Mean	Mode	Mean
N02A	2338	-	2337	32	2358	2338	-	2337	33	4666	4665	4589	4666	4665
N02B	2338	-	2337	32	2358	2338	-	2337	33	4666	4665	4589	4666	4665
N03A	2338	-	2337	32	2358	2338	-	2337	33	4666	4665	4589	4666	4665
N03B	1054	-	1053	14	1074	1054	-	1053	15	2098	2097	2075	2098	2097
N04A	2338	-	2337	32	903	2338	-	2337	33	4666	4665	1766	4666	4665
P02A	4320	-	4318	70	4339	4320	-	4319	5	4320	4320	4341	-	-
Q02A	4006	-	4006	0	18672	4006	-	4006	0	8008	8007	36511	8008	8008
Q02B	4006	-	4006	0	18672	4006	-	4006	0	8008	8007	36511	8008	8008
Q03A	4006	-	4006	0	20745	4006	-	4006	0	8008	8007	40385	8008	8007
Q03B	7910	4006	7909	23	41415	7910	4006	7909	27	15816	15815	80652	15816	15815
Q04B	7910	4006	7909	23	41415	7910	4006	7909	27	15816	15815	80652	15816	15815
R02A	3072	-	3072	0	2943	3072	-	3072	0	6144	6144	5765	6144	6144
R02B	9232	8	9226	229	9234	9232	8	9227	203	18456	18443	18066	18456	18447
R03A	3072	-	3072	0	2943	3072	-	3072	0	6144	6144	5764	6144	6144
R03B	9232	8	9231	56	9239	9232	8	9231	62	18456	18455	18078	18456	18455
R04B	9232	-	9232	0	9240	9232	8	9231	21	18456	18455	18079	18456	18455
S01B	1737	1162	1681	175	1702	43137	28762	41753	4392	3474	3358	3308	-	-
S02B	2895	2320	2837	186	2858	25895	20720	25371	1704	5790	5670	5571	-	-
S03A	579	-	574	52	593	7479	-	7422	648	1158	1145	1142	14958	14834
S04A	579	-	579	0	598	7479	-	7479	0	1158	1158	1154	14958	14958
S05A	579	-	570	68	2116	7479	-	7356	948	1158	1132	3560	14958	14641
T01A	1026	-	1026	0	1046	1026	-	1026	0	2052	2052	2030	2052	2052
T01B	1026	-	1026	0	1046	1026	-	1026	0	2052	2052	2030	2052	2052
T02A	1026	-	1026	0	1046	1026	-	1026	0	2052	2052	2030	2052	2052
T02B	1026	-	1026	0	1046	1026	-	1026	0	2052	2052	2030	2052	2052
T03A	1026	-	1026	0	1046	1026	-	1026	0	2052	2052	2030	2052	2052
T03B	1026	-	1026	0	1046	1026	-	1026	0	2052	2052	2030	2052	2052
T04A	1026	-	1026	0	1807	1026	-	1026	0	2052	2052	3531	2052	2052
U01A	1056	-	1056	0	1076	1056	-	1056	0	2096	2096	2073	2096	2096
U01B	1056	-	1056	0	1076	1056	-	1056	0	2096	2096	2073	2096	2096
U02A	1056	-	1056	0	1076	1056	-	1056	0	2096	2096	2073	2096	2096
U03A	290	-	290	0	309	1056	-	1056	0	580	580	587	2096	2096
U03B	578	-	578	0	597	1056	-	1056	0	1156	1156	1152	2096	2096
U04A	290	-	290	0	309	1056	-	1056	0	580	580	587	2096	2096
U04B	578	-	578	0	597	1056	-	1056	0	1156	1156	1152	2096	2096
V01A	312	-	312	0	911	312	-	312	0	608	608	1529	608	608
V02A	3152	-	3151	11	6289	3152	-	3151	9	6288	6287	12059	6288	6287
V03A	5968	-	5968	0	11954	5968	-	5967	8	11920	11920	23136	11920	11919
V03B	18928	18888	16777	1933	33625	18928	18888	16858	1843	37840	33692	65767	37840	33685
V04A	257	-	257	0	943	257	-	257	0	514	514	1577	514	514
W01A	520	-	519	16	1033	1032	-	1030	33	1040	1038	2023	2064	2061
W01B	520	-	520	0	530	1032	-	1032	0	1040	1040	1029	2064	2064
W02A	520	-	520	0	530	1032	-	1032	0	1040	1040	1029	2064	2064
W02B	520	-	520	0	530	1032	-	1032	0	1040	1040	1029	2064	2064
W03A	520	-	520	0	530	1032	-	1032	0	1040	1040	1029	2064	2064
W04A	520	-	520	0	530	1032	-	1032	0	1040	1040	1029	2064	2064
W05A	520	-	520	0	530	1032	-	1032	0	1040	1040	1029	2064	2064
X02A	577	-	559	100	578	577	-	557	103	1154	1114	1110	1154	1117
X03A	577	-	573	43	592	577	-	573	47	1154	1146	1142	1154	1146
X04A	577	-	573	43	592	577	-	573	47	1154	1146	1142	1154	1146
Y02A	-	-	-	-	-	-	-	-	-	-	-	-	-	-
Y02B	-	-	-	-	-	-	-	-	-	-	-	-	-	-
Y03A	5199	-	5132	584	5153	-	-	-	-	10398	10271	10079	-	-
Y03B	10398	-	10383	392	10405	-	-	-	-	20796	20765	20353	-	-

Table 6: For group 0 SDKs received February to June 2011, Size of biometric feature data, for single-eye (left) and two-eye (right), and for enrollment and search templates. In each block the columns are: *Mode* the most common value - **yellow indicates this is the most important column**. *Second* the second most common value (0 for failure to make template, "-" for no occurrence), *Mean* is the arithmetic mean; *SD* is the standard deviation; *Disk* is the template size implied by dividing the size of the data resident on disk after post-enrollment finalization by the size of the enrolled population. Green cells indicate size less than or equal to 512 bytes. Most providers produce fixed length templates from each image (i.e. not random variables dependent on the data - some implementations from Q, S and V produce variable length templates). In any case, the values in each cell are the statistics for all templates not including template generation failures. Template size is not a function of enrolled population size, i.e. it is not tailored for particular values of N.

	Single Eye									Two eyes				
	Enrollment N = 160000					Search S = 553230				Enrollment N = 160000			Search S = 315662	
	Template				Disk	Template				Template		Disk	Template	
	Mode	Second	Mean	SD.	Mean	Mode	Second	Mean	SD.	Mode	Mean	Mean	Mode	Mean
N11A	2338	-	2337	32	903	2338	-	2337	33	4666	4665	1766	4666	4665
N11B	992	-	991	13	1011	992	-	991	14	1974	1973	1953	1974	1973
N12A	2338	-	2337	32	903	2338	-	2337	33	4666	4665	1766	4666	4665
N12B	2338	-	2328	146	2349	2338	-	2326	163	4666	4645	4570	4666	4644
N13B	2338	-	2328	146	2349	2338	-	2326	163	4666	4645	4570	4666	4644
P11A	4320	-	4082	983	4102	4320	-	4036	1069	8640	8121	7959	8640	8096
P11B	4320	-	4320	0	4341	4320	-	4320	0	8640	8640	8481	-	-
Q11B	7910	4006	7903	162	40866	7910	4006	7901	183	15816	15800	79583	15816	15800
Q12B	8422	4262	8414	173	8436	8422	4262	8412	195	16840	16823	16495	16840	16823
Q13B	4262	4006	4261	10	4282	4262	4006	4261	11	8520	8505	8350	8520	8505
R11A	3072	-	3072	0	2943	3072	-	3072	0	6144	6144	5764	6144	6144
R11B	9232	-	9232	0	9240	9232	-	9232	0	18456	18456	18079	18456	18456
R12A	3072	-	3072	0	2943	3072	-	3072	0	6144	6144	5764	6144	6144
S11A	579	-	579	0	1355	7479	-	7479	0	1158	1158	2370	14958	14958
S11B	2895	2320	2871	205	8168	25895	20720	25645	2010	5790	5733	15306	51790	51302
S12A	579	-	579	0	1355	7479	-	7479	0	1158	1158	2370	14958	14958
T11A	1028	-	1028	0	2068	1028	-	1028	0	2056	2056	4039	2056	2056
T11B	1028	-	1028	0	1048	1028	-	1028	0	2056	2056	2034	2056	2056
T12B	1028	-	1028	0	2068	1028	-	1028	0	2056	2056	4039	2056	2056
U11A	544	-	544	0	563	1056	-	1056	0	1072	1072	1070	2096	2096
U11B	1056	-	1056	0	1076	1056	-	1056	0	2096	2096	2073	2096	2096
U12A	1056	-	1056	0	1076	1056	-	1056	0	2096	2096	2073	2096	2096
U12B	1056	-	1056	0	1076	1056	-	1056	0	2096	2096	2073	2096	2096
V11A	3268	-	3268	0	6833	3268	-	3268	0	6520	6520	13109	6520	6520
V11B	20008	19968	17856	1934	35800	20008	19968	17937	1844	40000	35851	70026	40000	35844
V12A	16228	16188	14135	1739	28622	16228	16188	14264	1667	32440	28382	55931	32440	28508
V12B	19506	19466	17354	1934	34221	19984	19944	17913	1844	38996	34847	66917	39936	35780
W11A	1032	-	1032	0	1043	1032	-	1032	0	2064	2064	2032	2064	2064
W11B	520	-	520	0	530	1032	-	1032	0	1040	1040	1029	2064	2064
W12A	1032	-	1032	0	1043	25608	-	25608	0	2064	2064	2032	51216	51216
X11A	577	-	573	42	593	577	-	573	46	1154	1146	1143	1154	1147
X11B	577	-	573	42	593	577	-	573	46	1154	1146	1143	-	-

Table 7: For group 1 SDKs received in August 2011, Size of biometric feature data, for single-eye (left) and two-eye (right), and for enrollment and search templates. In each block the columns are: *Mode* the most common value - **yellow indicates this is the most important column**. *Second* the second most common value (0 for failure to make template, "-" for no occurrence), *Mean* is the arithmetic mean; *SD* is the standard deviation; *Disk* is the template size implied by dividing the size of the data resident on disk after post-enrollment finalization by the size of the enrolled population. Green cells indicate size less than or equal to 512 bytes. Most providers produce fixed length templates from each image (i.e. not random variables dependent on the data - some implementations from Q, S and V produce variable length templates). In any case, the values in each cell are the statistics for all templates not including template generation failures. Template size is not a function of enrolled population size, i.e. it is not tailored for particular values of N.

4.2 Exploiting multiple cores

The IREX III API, CONOPS, AND EVALUATION SPECIFICATION document did not support execution of a search across $B > 1$ blades because there was no need. The reasons are as follows:

▷ In all cases, for all population sizes, the entire enrollment database is small enough to fit in main memory.

▷ A blade equipped with $C > 1$ cores was fully utlized by running C searches simultaneously as separate processes. This facility was not used for threaded implementations. (Timing measurements were made with $C = 1$ process).

▷ When searching an enrollment database of size E on a blade with memory M, the number of copies of the enrollment data that can be made and kept in memory is $c = \lfloor M/E \rfloor$. This supports execution of $\min(c, C)$ completely independent processes, each running separate searches.

▷ However, we can avoid this memory limit by making only $c = 1$ copies of the enrollment database by using the LINUX *fork()* system call C times. While this spawns C entirely separate processes, the LINUX implementation of *fork()* uses *copy-on-write* semantics, which means that the enrollment data is not copied because it is read-only.

4.3 Comparing search times for threaded operations

Operationally, threading is used in biometric systems to expedite one-to-many search. Accordingly, for testing, IREX III allowed the search operation to be threaded - implementers were free to use threading or not. A few elected to do so, while most operated in single thread mode. In any case, timing estimates are made by wrapping the core *identify_template* function call in a high resolution timer. Then, to render comparisons, the search durations for threaded implementations should be adjusted to account for the number of computational threads used. The adjustment would take the form of a multiplier $\eta(N, K)$ such that the single-thread time was $T_1 = \eta(N, K)T_K$ for population size, N, and number of cores used, K. Initially the first order correction $\eta(N, K) = K$ was applied but IREX III participants noted that this was unrealistically punitive, because of Amdahl's law[18] and the fact that while the number of cores increases by K, other parts of the system architecture (e.g. memory pipes) do not.

One participant submitted identical algorithms, $U01A$ and $U02A$, in threaded ($K = 16$) and unthreaded ($K = 1$) versions. This afforded the following analysis and improved correction. Both algorithms executed 1000 searches in each of four populations N = 20,000, N= 160,000, N=1,600,000, N=3,904,239 on the IREX III API specified AMD processor-based machine. The results, shown in Figure 5 show that the threaded implementation achieves close to K=16 times speed improvement (15.3) for the smallest enrollment database, but this reduces rapidly with N (to 7.68). The functional form for

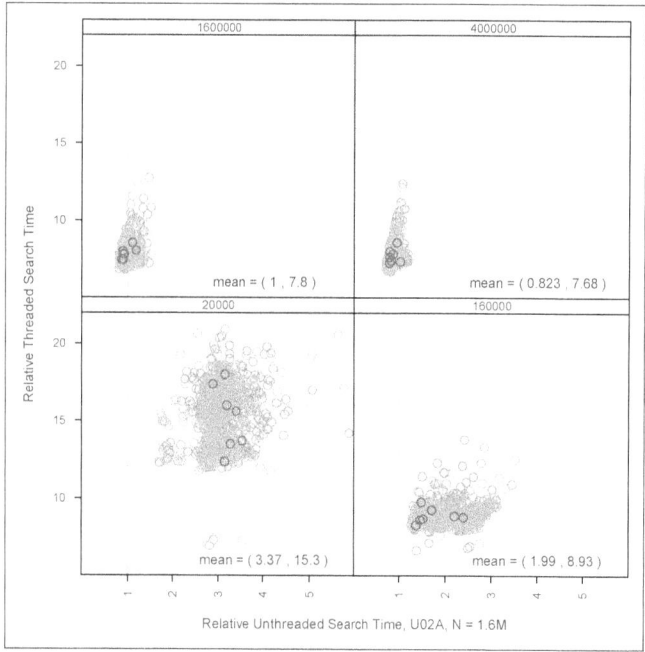

Figure 5: Comparison of the speed of threaded and unthreaded implementations of the same identification search algorithm. U01A here runs 16 threads, U02A runs 1. On the vertical axis is the U01A duration normalized by the duration of U02A searches on a population of N = 1,600,000. The horizontal axis shows the U02A durations, normalized by the same quantity. Each point corresponds to a search, with mate and nonmate searches colored separately. The four panels apply to the same 1000 searches in four enrolled population sizes, $20K \leq N \leq 3.9M$. The mean relative duration values show threaded speeds are multiplied by less than $K = 16$, with the factor decreasing with increasing N.

| FNIR = FALSE NEGATIVE IDENT. RATE | N = NEUROTECHNOLOGY | P = SMU | Q = IRITECH | R = COGENT | S = SMARTSENSORS | T = CAMBRIDGE |
| FPIR = FALSE POSITIVE IDENT. RATE | U = L1 | V = MORPHO | W = IRISID | X = CROSSMATCH | Y = KYNEN | |

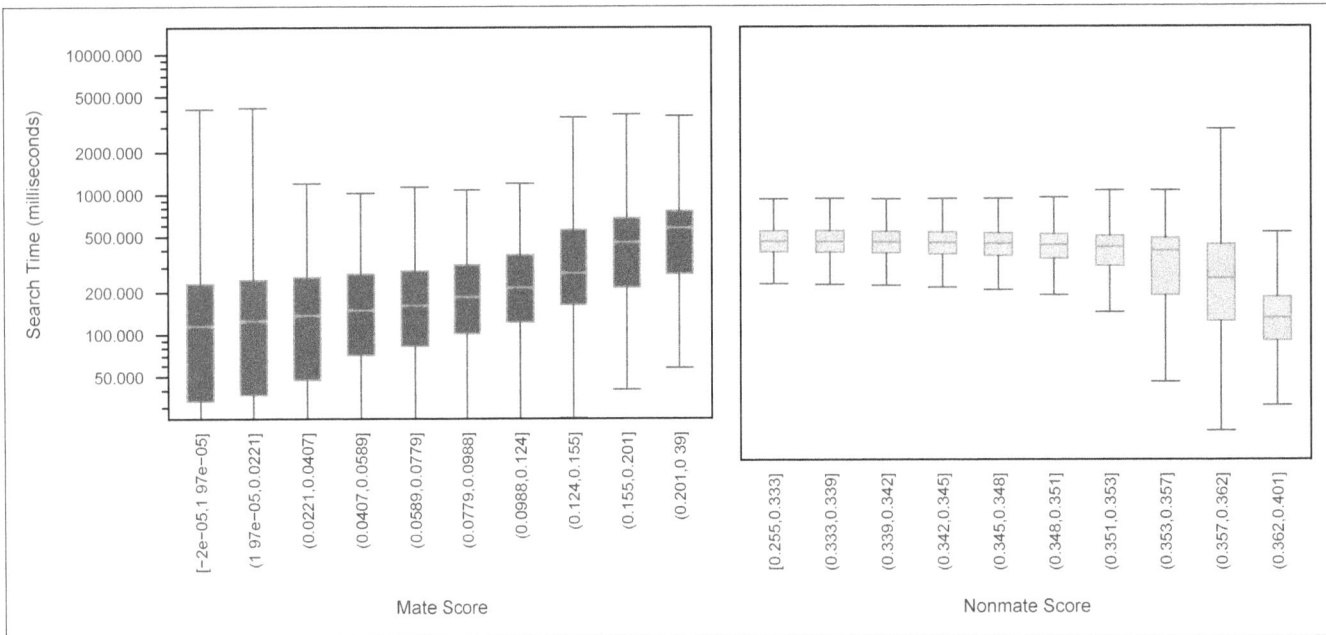

Figure 6: Dependence of speed and comparison score. For the fast T11A algorithm, the panels show the distribution of the one-to-many search duration for ten equal-width intervals of the mate dissimilarities, and separately for the nonmate dissimilarities too. For many algorithms, search duration is independent of the eventual mate score. Here, the downward trend at left shows that search is faster when the ultimate mate score is low (i.e. good). This indicates that, for this algorithm, search will be faster if all factors that influence outcome are benign. Similar figures for all algorithms appear in the Appendices.

$\eta(N, K)$ is not evident from this data, but various forms have been documented elsewhere [18] and implementers should consider the issue.

NB: In all figures and tables that follow, these corrections were not applied to the threaded implementation durations. Instead a fixed factor of 16 was used. This disadvantages threaded implementations by a factor ranging from nearly 1 (16/15.3) up to about 2 (16/7.8), depending on N. This was sustainable because most algorithms were submitted as single threaded implementations.

4.4 Data dependence

Given a fixed length template and an elemental one-to-one comparison duration η, the naïve estimate of a search in an N person population would be $T = \eta N$. This would be constant, fixed for all searches. However, several effects render this an incorrect model. Some of these are due to the non-ideal nature of the system under test, while the larger more important ones are due to the use of several fast search methods implemented within the SDK. While these are intended to expedite search, they can introduce a dependency on the data itself.

▷ When comparing two templates, the SDK implements search over in-plane rotation of the iris until a low dissimilarity is found or some angular limit is reached.

▷ When comparing two templates, the SDK aborts the dissimilarity computation when dissimilarity has accumulated beyond some limit (i.e. early declaration that score is a nonmate score).

▷ The IREX III API supports the construction of an index from the enrollment data. Indexing schemes have been described in the literature[16, 15], and promise expeditious search.

▷ In an N-enrollee population, the SDK aborts after a low-dissimilarity is found so that $n \leq N$. The duration then depends on the position of the mate in the population. This strategy is contraindicated in most operations where more than one mate could be present in the database. In real-valued template representations, this process can be expedited if features are arranged to be ordered by decreasing variance. In binary template representations, higher entropy bits could be ordered first.

▷ A search produces the k nearest neighbors (according to some dissimilarity or distance metric). Searches can be expedited if a template comparison is discontinued at the point the partial dissimilarity sum exceeds the k-th closest element. There is an extensive literature on finding k nearest neighbors (for general pattern recognition tasks); see for example kd-trees[10].

Figure 6 shows an example where search speed and recognition result are correlated. For algorithm T11A, the duration of a search drops with the dissimilarity score ultimately produced by that search. This may be a by-product of an indexing scheme.

4.5 Results

Template Generation: Table 8 gives median template generation times. These give the time needed to convert an input image[13] to a template. All operations are in memory, so the estimates do not include I/O. Figure 9 gives boxplots of the search template generation durations. Within each box, the bar indicates the median, and vertical edges indicate the inter-quartile range (IQR), and the whiskers extend to 1.5IQR to cover approximately 95% of normally distributed data. Outliers are present but not shown.

Regarding template generation, the notable observations are as follows.

▷ Template generation times span a factor of at least forty, ranging from below 20 to above 800 milliseconds. The faster algorithms (most notably from W, and also from T, U, S and Q (class A) implementations) run in tens of milliseconds and these could therefore be readily produced in embedded devices such as handheld and mobile iris cameras. This speed is also valuable during capture when several images of an eye are collected and template matching or quality analysis is used to select the best image.

▷ Some class A implementations exceed the IREX III specification that 90% of template generations should take fewer than 0.5 seconds (the vertical line in Figure 9). This limit is not enforced here because these implementations should be relabeled and accepted as class B.

▷ The variance in template generation times also varies between algorithms. Low variance is often a desirable property and while some algorithms run in close to constant time, variance in fast implementations is probably more tolerable than in slow ones.

▷ As evident in the template size results in Tables 6 and 7 a few implementations use asymmetric templates: that is the search template is different from, yet matchable with, the enrollment template. This results in different generation times most notably for W12A, but also for S01B, S02B, S11B, V03B, Y03A, and Y03B.

Search: Tables 9 and 10 show the duration of the one-to-many identification implementations. These are the times needed to search a template against the N in-memory enrolled identities. The Tables show speed for four population sizes from 20,000 to the nominal 4M cap (actually 3.9M - see section 3.1). The Tables also show the implied number of comparisons per second obtained by dividing the enrolled population size by the duration.

Regarding search, the notable observations are as follows.

---

[13]Mostly 640x480 images, but some 480x480 also.

(a) Group 0			(b) Group 1		
SDK	Template Gen. Time		SDK	Template Gen. Time	
	Enrol	Search		Enrol	Search
N02A	456 ± 4	460 ± 4	N11A	446 ± 3	447 ± 3
N02B	453 ± 4	460 ± 3	N11B	447 ± 3	447 ± 3
N03A	453 ± 4	452 ± 3	N12A	448 ± 3	450 ± 3
N03B	441 ± 4	446 ± 3	N12B	444 ± 4	447 ± 4
N04A	444 ± 3	445 ± 3	N13B	447 ± 4	448 ± 3
P02A	117 ± 1	122 ± 0	P11A	384 ± 3	388 ± 3
Q02A	27 ± 0	32 ± 0	P11B	2880 ± 27	2146 ± 18
Q02B	27 ± 0	29 ± 0	Q11B	444 ± 0	460 ± 0
Q03A	82 ± 0	89 ± 0	Q12B	443 ± 0	458 ± 0
Q03B	214 ± 0	226 ± 0	Q13B	258 ± 0	258 ± 0
Q04B	213 ± 0	221 ± 0	R11A	561 ± 2	564 ± 2
R02A	518 ± 2	530 ± 2	R11B	669 ± 3	708 ± 3
R02B	506 ± 4	530 ± 1	R12A	562 ± 2	563 ± 2
R03A	536 ± 6	541 ± 2	S11A	65 ± 4	89 ± 4
R03B	502 ± 1	517 ± 1	S11B	458 ± 6	516 ± 6
R04B	479 ± 1	491 ± 1	S12A	63 ± 4	79 ± 4
S01B	714 ± 12	859 ± 12	T11A	53 ± 0	56 ± 0
S02B	683 ± 11	765 ± 11	T11B	56 ± 0	56 ± 0
S03A	62 ± 6	85 ± 6	T12B	56 ± 0	54 ± 0
S04A	60 ± 6	82 ± 6	U11A	59 ± 1	59 ± 1
S05A	61 ± 6	87 ± 7	U11B	163 ± 1	168 ± 1
T01A	65 ± 0	70 ± 0	U12A	164 ± 1	169 ± 1
T01B	65 ± 0	66 ± 0	U12B	163 ± 1	169 ± 1
T02A	31 ± 0	30 ± 0	V11A	427 ± 1	431 ± 1
T02B	65 ± 0	66 ± 0	V11B	719 ± 2	731 ± 2
T03A	38 ± 0	40 ± 0	V12A	472 ± 1	482 ± 1
T03B	38 ± 0	39 ± 0	V12B	665 ± 2	677 ± 2
T04A	38 ± 0	39 ± 0	W11A	29 ± 0	28 ± 0
U01A	56 ± 1	57 ± 1	W11B	28 ± 0	29 ± 0
U01B	52 ± 1	57 ± 1	W12A	25 ± 0	307 ± 1
U02A	53 ± 1	61 ± 1	X11A	372 ± 3	370 ± 3
U03A	52 ± 1	57 ± 1	X11B	372 ± 3	372 ± 2
U03B	51 ± 1	54 ± 1			
U04A	53 ± 1	55 ± 1			
U04B	51 ± 1	57 ± 1			
V01A	246 ± 0	254 ± 0			
V02A	688 ± 1	705 ± 1			
V03A	578 ± 1	583 ± 1			
V03B	757 ± 1	625 ± 1			
V04A	283 ± 0	287 ± 0			
W01A	34 ± 0	20 ± 0			
W01B	20 ± 0	20 ± 0			
W02A	19 ± 0	27 ± 0			
W02B	18 ± 0	22 ± 0			
W03A	19 ± 0	20 ± 0			
W04A	19 ± 0	26 ± 0			
W05A	19 ± 0	20 ± 0			
X02A	290 ± 2	286 ± 2			
X03A	262 ± 2	260 ± 2			
X04A	263 ± 2	263 ± 2			
Y02B	102 ± 1	68 ± 0			
Y03A	84 ± 0	54 ± 0			
Y03B	189 ± 1	121 ± 1			

Table 8: For group 0 implementations, submitted February to June, 2011, and for group 1 implementations, submitted August, 2011, the duration, in milliseconds, of calls to the image-to-template conversion function. Measurements are given for both initial enrollment and for preparation of a template for one-to-many search. The template generation times are medians estimated over 1000 template generations executed on the IREX III API-specified LINUX blade and for four different enrolled population sizes (i.e. 8,000 measurements in total). The uncertainty estimates are two-sided 95% confidence intervals computed from ordinary bootstrap resampling of the measured durations. Light and dark green colors indicate durations less than or equal to 200 and 20 milliseconds respectively.

▷ The matches-per-second values range from below 10,000 to beyond 8,000,000. This three-order-of-magnitude range is indicative of radically different algorithms, both in terms of core 1:1 template comparison and fast large population search. Two providers, T and N produce several algorithms exceeding 5 million matches per second while some algorithms from S, U, and W exceed 2 million matches per second. These are joined by R and V at the 1 million mark.

▷ Several implementers submitted algorithms varying greatly in speed, with two orders of magnitude applying to all except T and U which are fast, and P, X and Y which are slower.

▷ Figure 12 plots an accuracy measure (FNIR at FPIR = 0.0001) versus search duration that usually scales linearly with the number of enrolled identities. That is, in a power law model $T = aN^b$, the value b is close to 1. The value of b is encoded using line color.

▷ Part of the variation in search speed can be attributed to the range of rotation values over which template comparison is done. While a wide range is needed for the single-eye legacy cameras used for this dataset, a narrower angular range should be viable for images collected exclusively with modern two-eye cameras. Search time grows linearly with angular range. While search can be terminated early if a low candidate mate score is found, almost all (i.e. at least N-1) comparisons are nonmated which requires search of the full angular range.

Total duration: Figures 7 and 8 show the dependency of three durations on population size N. These are template generation time; search time; and the sum of those two. The graphs show that search time is not always the dominant component of total search duration.

▷ The duration of the template generation process is independent of N for all processes. This is the expected result - some developers could, in principle, adopt a different representation to handle larger population sizes, and this would lead to a population size dependency.

▷ For most implementations the template generation time is less than the search time - these values have been tabulated previously in Tables 8, 9 and 9. However for some algorithms the search time is faster and the template generation time is a substantial component of the total until N reaches very high values. Thus for algorithm N04A, N12A the search duration only exceeds template generation time for N above one million. For other algorithms, this break even point is often in the tens or hundreds of thousands.

This is important for "smaller" applications where search might not be spread across several cores. For example, the computation time needed for a one-to-many gymnasium access control system to identify one of a few thousand members would be dominated by template generation time. Practically, image capture and network transmission times would also be large components of the total.

▷ Note that gains of a factor of two or three in duration that might be realized by, for example, compiler optimization or partitioning the enrollment database by sex or by eye-label, are substantially smaller than the range of search times across the algorithms measured here.

4.6   Conclusion

Figures 9, 10 and 11 show, respectively, distributions of the durations of template generation, search, and their sum. The first two reveal a factor of 40 between the fastest and slowest template generation algorithms, and a factor of 1000 for search. These variations belie the reputation, arising in the academic literature for Daugman-like algorithms, that iris is (always) fast. A better statement is that it is often fast (as with most Cambridge algorithms), but sometimes is not.

FNIR = FALSE NEGATIVE IDENT. RATE	N = NEUROTECHNOLOGY	P = SMU	Q = IRITECH	R = COGENT	S = SMARTSENSORS	T = CAMBRIDGE
FPIR = FALSE POSITIVE IDENT. RATE	U = L1	V = MORPHO	W = IRISID	X = CROSSMATCH	Y = KYNEN	

SDK	Num.	One-to-Many Search Time				Implied Comparisons per Second			
	Threads	N = 20000	N = 160000	N = 1600000	N = 3904239	N = 20000	N = 160000	N = 1600000	N = 3904239
N02A	1	667 ± 1	5350 ± 10	53800 ± 103	129000 ± 248	3.00E+04	2.99E+04	2.98E+04	3.17E+04
N02B	1	113 ± 2	900 ± 16	9050 ± 164	22300 ± 404	1.78E+05	1.78E+05	1.77E+05	1.84E+05
N03A	1	265 ± 2	2130 ± 21	21300 ± 209	51800 ± 510	7.56E+04	7.53E+04	7.50E+04	7.92E+04
N03B	1	126 ± 0	1020 ± 7	10000 ± 69	24400 ± 169	1.59E+05	1.57E+05	1.60E+05	1.68E+05
N04A	1	6 ± 0	31 ± 0	292 ± 4	708 ± 11	3.20E+06	5.07E+06	5.47E+06	5.79E+06
P02A	16	3880 ± 126	24500 ± 780	250000 ± 6938	621000 ± 17180	5.15E+03	6.53E+03	6.40E+03	6.60E+03
Q02A	1	124 ± 5	963 ± 45	9790 ± 455	24400 ± 1128	1.61E+05	1.66E+05	1.64E+05	1.68E+05
Q02B	1	119 ± 5	957 ± 44	9730 ± 453	24300 ± 1120	1.68E+05	1.67E+05	1.64E+05	1.69E+05
Q03A	1	43 ± 2	264 ± 23	2850 ± 248	7510 ± 624	4.56E+05	6.06E+05	5.62E+05	5.46E+05
Q03B	1	493 ± 30	3340 ± 243	34500 ± 2485	87100 ± 6120	4.06E+04	4.79E+04	4.64E+04	4.71E+04
Q04B	1	1410 ± 57	10800 ± 459	110000 ± 4669	270000 ± 11380	1.42E+04	1.48E+04	1.45E+04	1.52E+04
R02A	1	23 ± 0	160 ± 0	1610 ± 1	3910 ± 3	8.43E+05	9.98E+05	9.93E+05	1.05E+06
R02B	8	3320 ± 20	28500 ± 167	281000 ± 1690	697000 ± 4175	6.03E+03	5.62E+03	5.69E+03	5.88E+03
R03A	1	23 ± 0	153 ± 0	1530 ± 1	3870 ± 4	8.61E+05	1.05E+06	1.04E+06	1.06E+06
R03B	1	103 ± 0	868 ± 0	9780 ± 10	26700 ± 31	1.95E+05	1.84E+05	1.64E+05	1.54E+05
R04B	1	197 ± 1	1590 ± 10	16900 ± 106	40900 ± 260	1.02E+05	1.01E+05	9.48E+04	1.00E+05
S01B	1	2960 ± 37	24400 ± 319	241000 ± 3158	577000 ± 7519	6.76E+03	6.55E+03	6.64E+03	7.10E+03
S02B	1	3380 ± 25	27100 ± 207	269000 ± 2074	663000 ± 5023	5.92E+03	5.91E+03	5.94E+03	6.18E+03
S03A	1	161 ± 1	1240 ± 12	12400 ± 122	30200 ± 299	1.24E+05	1.29E+05	1.29E+05	1.36E+05
S04A	1	155 ± 2	1230 ± 19	12300 ± 191	30100 ± 469	1.29E+05	1.30E+05	1.30E+05	1.36E+05
S05A	1	55 ± 0	180 ± 3	1060 ± 21	2820 ± 59	3.64E+05	8.90E+05	1.50E+06	1.45E+06
T01A	1	32 ± 0	282 ± 1	2790 ± 15	6020 ± 36	6.13E+05	5.68E+05	5.75E+05	6.80E+05
T01B	1	529 ± 0	4210 ± 1	42600 ± 16	103000 ± 32	3.78E+04	3.80E+04	3.76E+04	3.98E+04
T02A	1	27 ± 0	217 ± 1	2150 ± 10	6030 ± 26	7.28E+05	7.38E+05	7.43E+05	6.79E+05
T02B	1	44 ± 0	355 ± 1	3530 ± 14	9370 ± 35	4.55E+05	4.51E+05	4.53E+05	4.37E+05
T03A	1	47 ± 0	378 ± 1	3790 ± 12	9270 ± 29	4.24E+05	4.23E+05	4.22E+05	4.42E+05
T03B	1	71 ± 0	573 ± 1	5710 ± 15	15000 ± 38	2.79E+05	2.79E+05	2.80E+05	2.72E+05
T04A	1	2 ± 0	14 ± 0	164 ± 0	492 ± 1	9.66E+06	1.13E+07	9.78E+06	8.33E+06
U01A	16	156 ± 9	812 ± 23	4040 ± 61	8420 ± 91	1.28E+05	1.97E+05	3.96E+05	4.87E+05
U01B	16	208 ± 3	1060 ± 28	4920 ± 53	10400 ± 115	9.63E+04	1.50E+05	3.25E+05	3.92E+05
U02A	1	51 ± 0	232 ± 1	2010 ± 17	4970 ± 40	3.92E+05	6.91E+05	7.95E+05	8.25E+05
U03A	1	14 ± 0	103 ± 0	754 ± 5	1770 ± 12	1.35E+06	1.56E+06	2.12E+06	2.31E+06
U03B	1	38 ± 0	186 ± 1	1530 ± 12	3650 ± 29	5.22E+05	8.59E+05	1.05E+06	1.12E+06
U04A	1	15 ± 0	103 ± 0	769 ± 5	1770 ± 13	1.33E+06	1.55E+06	2.08E+06	2.32E+06
U04B	1	37 ± 0	186 ± 1	1530 ± 12	3650 ± 30	5.27E+05	8.60E+05	1.05E+06	1.12E+06
V01A	16	157 ± 3	663 ± 18	2770 ± 46	4700 ± 70	1.27E+05	2.42E+05	5.78E+05	8.72E+05
V02A	16	574 ± 10	1460 ± 30	5670 ± 88	5320 ± 61	3.48E+04	1.09E+05	2.82E+05	7.70E+05
V03A	1	130 ± 1	964 ± 7	9730 ± 79	23800 ± 190	1.54E+05	1.66E+05	1.65E+05	1.72E+05
V03B	1	142 ± 3	1060 ± 17	10200 ± 79	24300 ± 188	1.41E+05	1.51E+05	1.57E+05	1.69E+05
V04A	1	18 ± 0	113 ± 0	1100 ± 2	2990 ± 8	1.11E+06	1.41E+06	1.45E+06	1.37E+06
W01A	1	60 ± 0	-	4850 ± 34	-	3.33E+05		-	3.30E+05	-
W01B	1	4 ± 0	39 ± 0	394 ± 2	956 ± 6	4.04E+06	4.02E+06	4.06E+06	4.29E+06
W02A	16	298 ± 3	934 ± 27	5630 ± 25	12600 ± 57	6.72E+04	1.71E+05	2.84E+05	3.24E+05
W02B	16	21 ± 0	199 ± 1	1640 ± 12	2820 ± 31	9.14E+05	8.05E+05	9.74E+05	1.46E+06
W03A	1	61 ± 0	469 ± 1	4550 ± 9	11000 ± 19	3.24E+05	3.41E+05	3.52E+05	3.74E+05
W04A	16	75 ± 2	573 ± 9	2880 ± 28	6610 ± 77	2.64E+05	2.79E+05	5.56E+05	6.20E+05
W05A	1	25 ± 0	196 ± 0	1910 ± 2	4890 ± 4	7.96E+05	8.18E+05	8.37E+05	8.39E+05
X02A	1	121 ± 25	994 ± 204	9430 ± 2033	22900 ± 4932	1.66E+05	1.61E+05	1.70E+05	1.79E+05
X03A	1	119 ± 26	1010 ± 229	9800 ± 2059	24100 ± 5037	1.68E+05	1.58E+05	1.63E+05	1.70E+05
X04A	1	1990 ± 14	15900 ± 111	161000 ± 1131	392000 ± 2760	1.01E+04	1.01E+04	9.96E+03	1.05E+04
Y02B	1	722 ± 16	-	-	-	2.77E+04	-	-	-
Y03A	1	789 ± 7	6270 ± 62	37400 ± 371	-	2.53E+04	2.55E+04	4.27E+04	-
Y03B	1	3130 ± 15	25100 ± 126	68300 ± 343	-	6.39E+03	6.37E+03	2.34E+04	-

Table 9: For group 0 implementations, submitted February to June, 2011, the duration, in milliseconds, of calls to the template-to-candidate list identification function by SDK, and population size, and, at right, the effective number of one-to-one matches per second. The times are medians estimated over 1000 searches executed on the IREX III API-specified LINUX blade. The uncertainty estimates are two-sided 95% confidence intervals computed from ordinary bootstrap resampling of the measured durations. The number of nonmate searches is 512, the number of mate searches is 488. Light and dark green shading indicates that the search speed exceeds 500,000 and 5,000,000 matches-per-second respectively. Pink, red and dark red shading indicate that the SDK violated the maximum duration limits by factors of 1, 3 and 10 respectively. In any given row, the number of matches per second is not constant due to random error and systematic memory and bus speed limitations of the PC architecture.

SDK	Num. Threads	One-to-Many Search Time				Implied Comparisons per Second			
		N = 20000	N = 160000	N = 1600000	N = 3904239	N = 20000	N = 160000	N = 1600000	N = 3904239
N11A	1	6 ± 0	32 ± 0	303 ± 4	732 ± 11	3.11E+06	4.94E+06	5.29E+06	5.60E+06
N11B	1	87 ± 0	707 ± 5	6930 ± 56	17000 ± 138	2.28E+05	2.26E+05	2.31E+05	2.41E+05
N12A	1	4 ± 0	21 ± 0	203 ± 3	492 ± 8	4.80E+06	7.36E+06	7.88E+06	8.32E+06
N12B	1	547 ± 5	4380 ± 45	43800 ± 454	109000 ± 1137	3.66E+04	3.66E+04	3.65E+04	3.75E+04
N13B	1	107 ± 2	855 ± 18	8550 ± 184	21100 ± 447	1.87E+05	1.87E+05	1.87E+06	1.94E+05
P11A	16	3410 ± 107	21800 ± 713	215000 ± 5964	539000 ± 14696	5.87E+03	7.33E+03	7.44E+03	7.61E+03
P11B	16	-	61900 ± 247	645000 ± 182		-	2.58E+03	2.48E+03	-
Q11B	1	432 ± 25	3090 ± 202	32100 ± 2077	80500 ± 5069	4.63E+04	5.17E+04	4.98E+04	5.09E+04
Q12B	1	1080 ± 25	8600 ± 204	88400 ± 2097	214000 ± 5068	1.86E+04	1.86E+04	1.81E+04	1.92E+04
Q13B	1	346 ± 6	2760 ± 55	27900 ± 560	69400 ± 1379	5.77E+04	5.79E+04	5.74E+04	5.90E+04
R11A	1	14 ± 0	91 ± 0	912 ± 1	2230 ± 3	1.39E+06	1.75E+06	1.75E+06	1.84E+06
R11B	1	308 ± 2	2450 ± 20	24800 ± 210	61200 ± 514	6.49E+04	6.52E+04	6.45E+04	6.70E+04
R12A	1	23 ± 0	153 ± 0	1540 ± 1	3910 ± 4	8.57E+05	1.04E+06	1.04E+06	1.05E+06
S11A	1	21 ± 0	77 ± 2	616 ± 20	1580 ± 51	9.52E+05	2.07E+06	2.60E+06	2.59E+06
S11B	1	393 ± 10	1850 ± 73	18000 ± 687	46300 ± 1708	5.09E+04	8.67E+04	8.87E+04	8.85E+04
S12A	1	22 ± 0	80 ± 2	622 ± 20	1590 ± 52	8.97E+05	1.98E+06	2.57E+06	2.58E+06
T11A	1	4 ± 1	35 ± 9	419 ± 100	969 ± 233	4.22E+06	4.54E+06	3.82E+06	4.23E+06
T11B	1	39 ± 1	332 ± 1	3120 ± 9	7640 ± 23	5.09E+05	4.83E+05	5.14E+05	5.36E+05
T12B	1	42 ± 0	339 ± 0	3490 ± 35	8470 ± 127	4.70E+05	4.73E+05	4.59E+05	4.84E+05
U11A	1	17 ± 0	116 ± 0	920 ± 4	2180 ± 10	1.13E+06	1.37E+06	1.74E+06	1.88E+06
U11B	1	84 ± 0	497 ± 3	4660 ± 28	11700 ± 79	2.38E+05	3.22E+05	3.43E+05	3.51E+05
U12A	1	62 ± 0	332 ± 1	3020 ± 16	7330 ± 38	3.21E+05	4.82E+05	5.30E+05	5.59E+05
U12B	1	95 ± 0	582 ± 3	4750 ± 36	13200 ± 68	2.09E+05	2.75E+05	3.37E+05	3.10E+05
V11A	1	119 ± 0	931 ± 4	9280 ± 46	22600 ± 112	1.67E+05	1.72E+05	1.72E+05	1.82E+05
V11B	1	231 ± 3	1790 ± 24	17500 ± 150	42400 ± 365	8.67E+04	8.96E+04	9.12E+04	9.67E+04
V12A	1	137 ± 2	1030 ± 14	9860 ± 51	23400 ± 118	1.46E+05	1.55E+05	1.62E+05	1.75E+05
V12B	16	3190 ± 53	20200 ± 290	190000 ± 1348	446000 ± 3256	6.28E+03	7.93E+03	8.40E+03	9.18E+03
W11A	1	122 ± 0	928 ± 2	9030 ± 16	21900 ± 33	1.64E+05	1.72E+05	1.77E+05	1.87E+05
W11B	1	18 ± 0	140 ± 0	1390 ± 1	3390 ± 2	1.06E+06	1.15E+06	1.15E+06	1.21E+06
W12A	1	2890 ± 4	22600 ± 21	223000 ± 111	545000 ± 232	6.91E+03	7.08E+03	7.17E+03	7.52E+03
X11A	1	118 ± 24	931 ± 195	9580 ± 1970	-	1.70E+05	1.72E+05	1.67E+05	-
X11B	1	1960 ± 9	16000 ± 81	157000 ± 790	386000 ± 1970	1.02E+04	9.97E+03	1.02E+04	1.06E+04

Table 10: For group 1 implementations, submitted August, 2011, the duration, in milliseconds, of calls to the template-to-candidate list identification function, by SDK, and population size, and, at right, the effective number of one-to-one matches per second. The times are medians estimated over 1000 searches executed on the IREX III API-specified LINUX blade. The uncertainty estimates are two-sided 95% confidence intervals computed from ordinary bootstrap resampling of the measured durations. The number of nonmate searches is 512, the number of mate searches is 488. Light and dark green shading indicates that the search speed exceeds 500,000 and 5,000,000 matches-per-second respectively. Pink, red and dark red shading indicate that the SDK violated the maximum duration limits by factors of 1, 3 and 10 respectively. In any given row, the number of matches per second is not constant due to random error and systematic memory and bus speed limitations of the PC architecture.

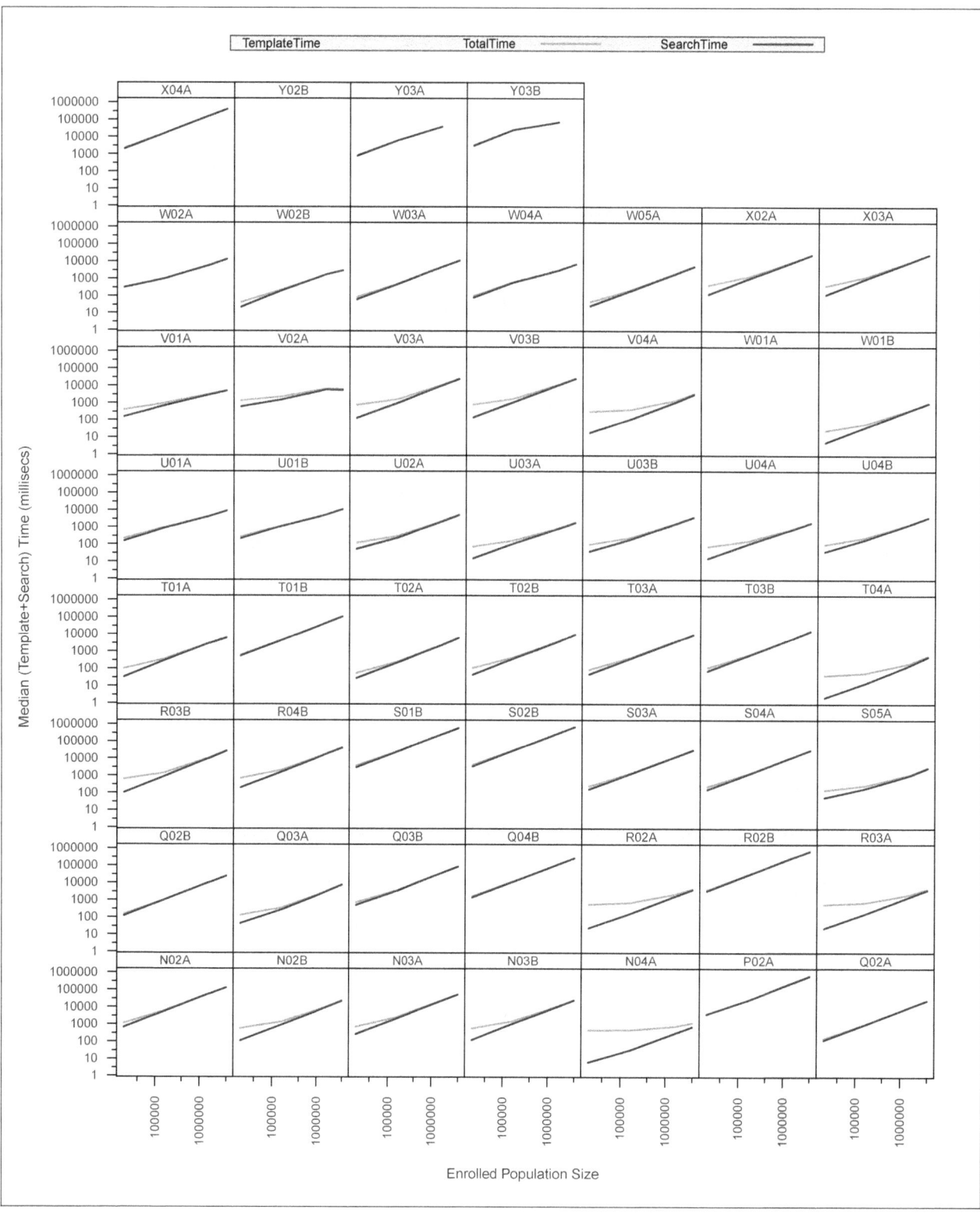

Figure 7: For group 0 implementations evaluated through June 2011, medians of the template generation, one-to-many search, and combined total durations as a function of enrolled population size. The times were measured on the IREX III API-specified LINUX blade. The statistics are estimated over 1000 samples executed on machines running one process. For implementations that use threading, the durations have been multiplied by the number of computational threads, but, as described in section 4.3, this disadvantges the few threaded implementations (see Tables 9 and 10). This multiplication is a first-order correction applied to make all durations comparable.

FNIR = FALSE NEGATIVE IDENT. RATE	N = NEUROTECHNOLOGY	P = SMU	Q = IRITECH	R = COGENT	S = SMARTSENSORS	T = CAMBRIDGE
FPIR = FALSE POSITIVE IDENT. RATE	U = L1	V = MORPHO	W = IRISID	X = CROSSMATCH	Y = KYNEN	

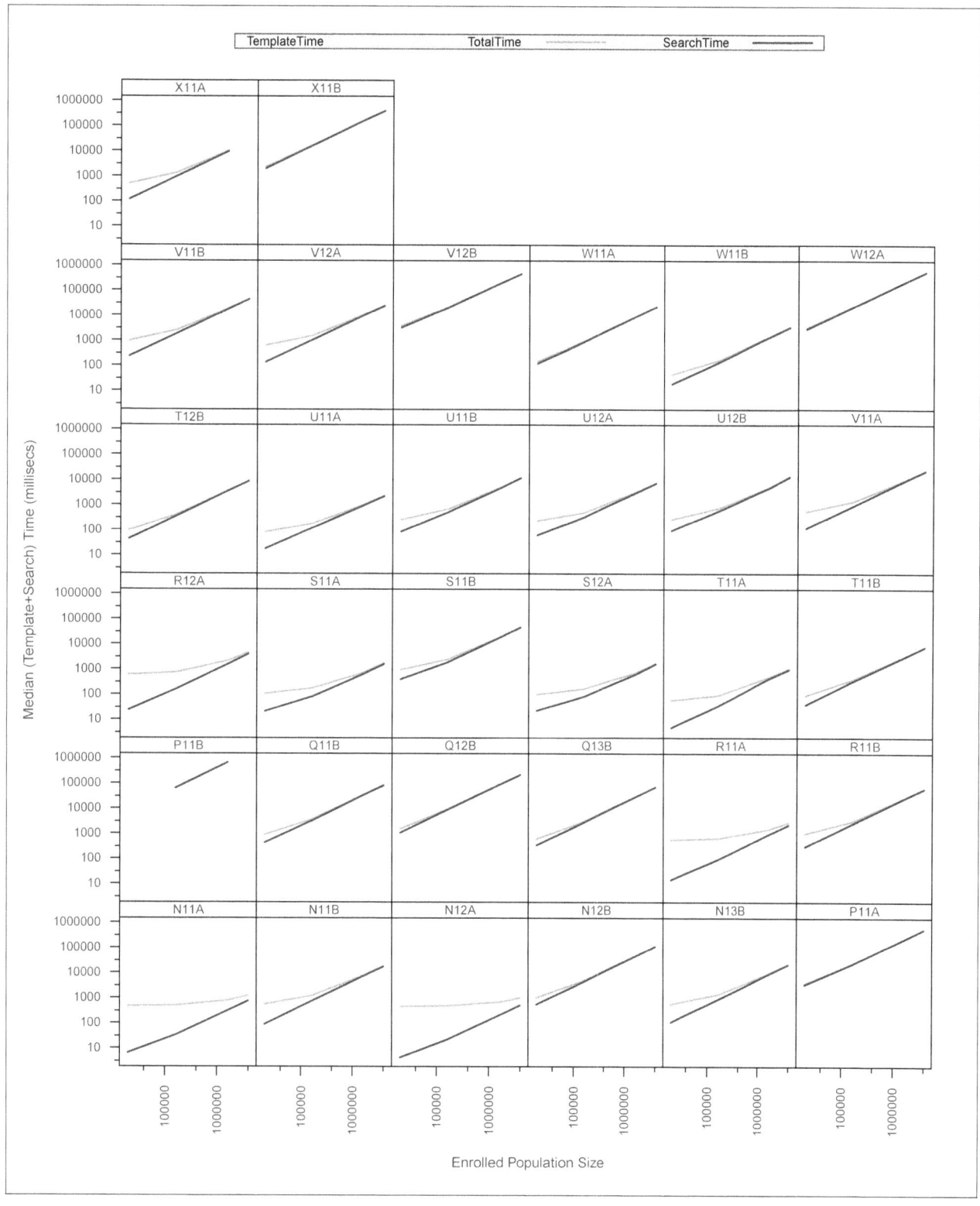

Figure 8: For group 1 implementations evaluated in August 2011, medians of the template generation, one-to-many search, and combined total durations as a function of enrolled population size. The times were measured on the IREX III API-specified LINUX blade. The statistics are estimated over 1000 samples executed on machines running one process. For implementations that use threading, the durations have been multiplied by the number of computational threads, but, as described in section 4.3, this disadvantges the few threaded implementations (see Tables 9 and 10). This multiplication is a first-order correction applied to make all durations comparable.

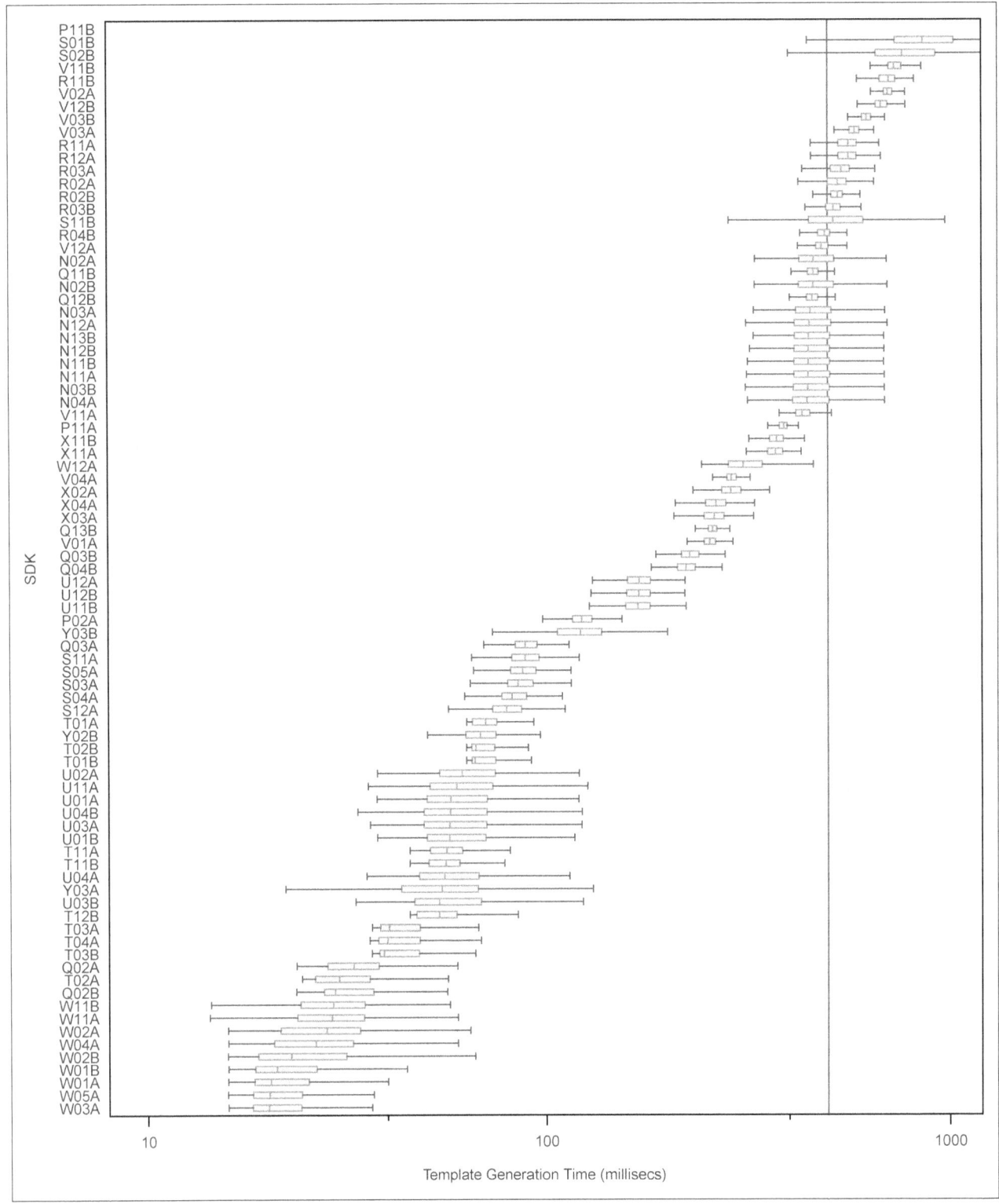

Figure 9: Distributions of template generation duration. The statistics are estimated over 1000 calls to the template generator used to make a search template on the IREX API-specified LINUX blade running one process. Template generation operations run without threading. The vertical line indicates the 90-th percentile IREX III time limits mandated for class A (fast, T ≤ 0.5s) SDKs. The line for class B (slow, T≤ 1.5s) SDKs is not visible.

| FNIR = FALSE NEGATIVE IDENT. RATE | N = NEUROTECHNOLOGY | P = SMU | Q = IRITECH | R = COGENT | S = SMARTSENSORS | T = CAMBRIDGE |
| FPIR = FALSE POSITIVE IDENT. RATE | U = L1 | V = MORPHO | W = IRISID | X = CROSSMATCH | Y = KYNEN | |

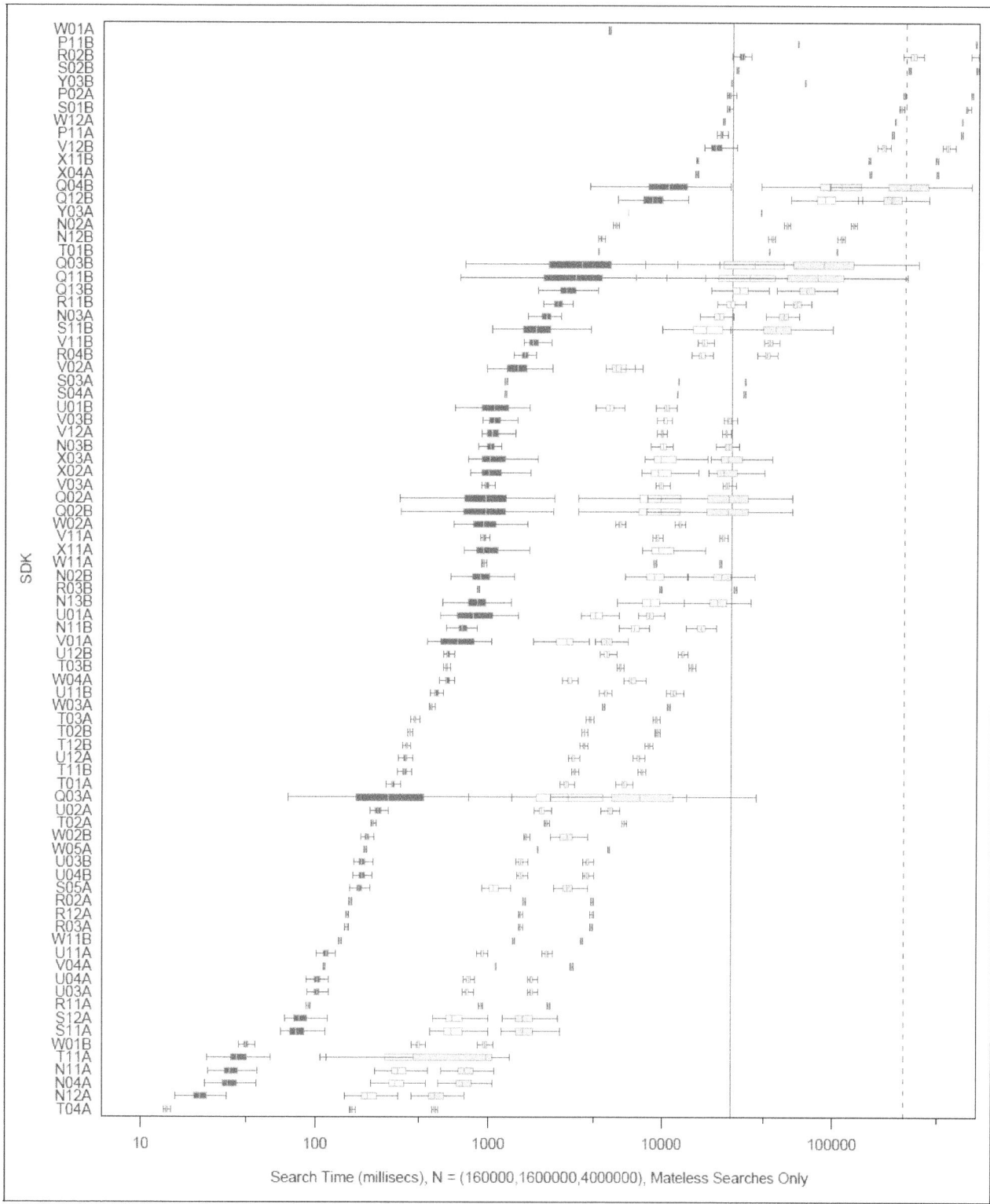

Figure 10: Distributions of search durations for enrolled population sizes of N=160,000 (blue), N=1,600,000 (gold) and N=3,904,239 (green). The statistics are estimated over 512 nonmate searches executed on the IREX API-specified LINUX blade running one process. For implementations that use threading, the durations have been multiplied by the number of computational threads; but, as described in section 4.3, this disadvantges the few threaded implementations (see Tables 9 and 10). This multiplication is a first-order correction applied to make all durations comparable. The vertical lines indicates the 90-th percentile IREX III time limits for class A (fast) and B (slow) SDKs for the three populations. All enrollments and all searches use a single iris image.

| FNIR = FALSE NEGATIVE IDENT. RATE | N = NEUROTECHNOLOGY | P = SMU | Q = IRITECH | R = COGENT | S = SMARTSENSORS | T = CAMBRIDGE |
| FPIR = FALSE POSITIVE IDENT. RATE | U = L1 | V = MORPHO | W = IRISID | X = CROSSMATCH | Y = KYNEN | |

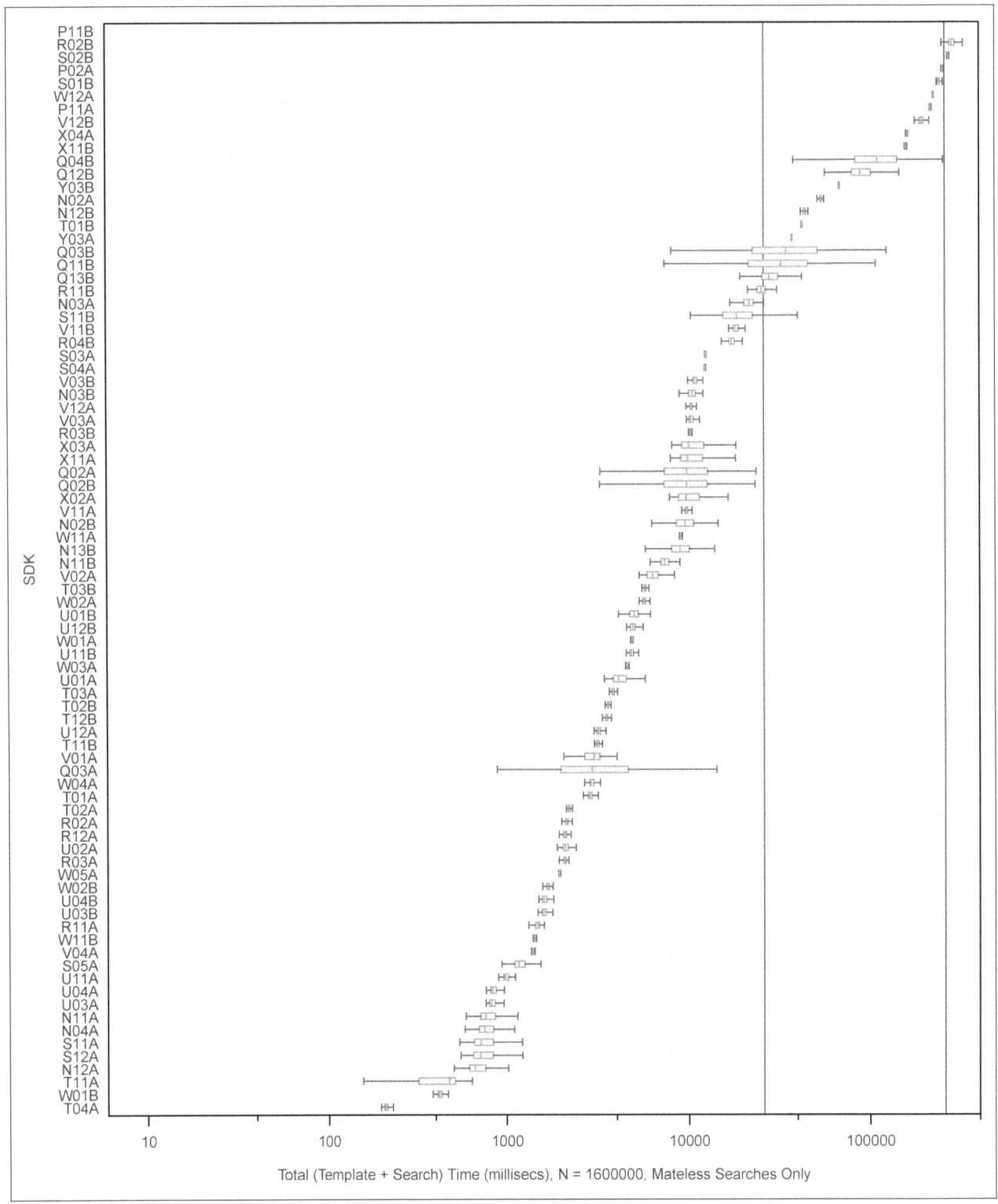

Figure 11: Distributions of combined search template generation plus one-to-many search durations, for N = 1,600,000, as estimated over 512 nonmate searches executed on the IREX III API-specified LINUX blade running one process. For threaded implementations (see Table 9), the search component has been multiplied by the number of computational threads. but, as described in section 4.3, this disadvantges the few threaded implementations (see Tables 9 and 10). This multiplication is a zero-order correction applied to make all durations comparable. Template generation is not threaded. As search time is dependent on N, and template generation time is not, the search time will dominate for large enough N - see Figures 7 and 8. The vertical lines indicate the IREX III time limits for class A and B SDKs.

| FNIR = FALSE NEGATIVE IDENT. RATE | N = NEUROTECHNOLOGY | P = SMU | Q = IRITECH | R = COGENT | S = SMARTSENSORS | T = CAMBRIDGE |
| FPIR = FALSE POSITIVE IDENT. RATE | U = L1 | V = MORPHO | W = IRISID | X = CROSSMATCH | Y = KYNEN | |

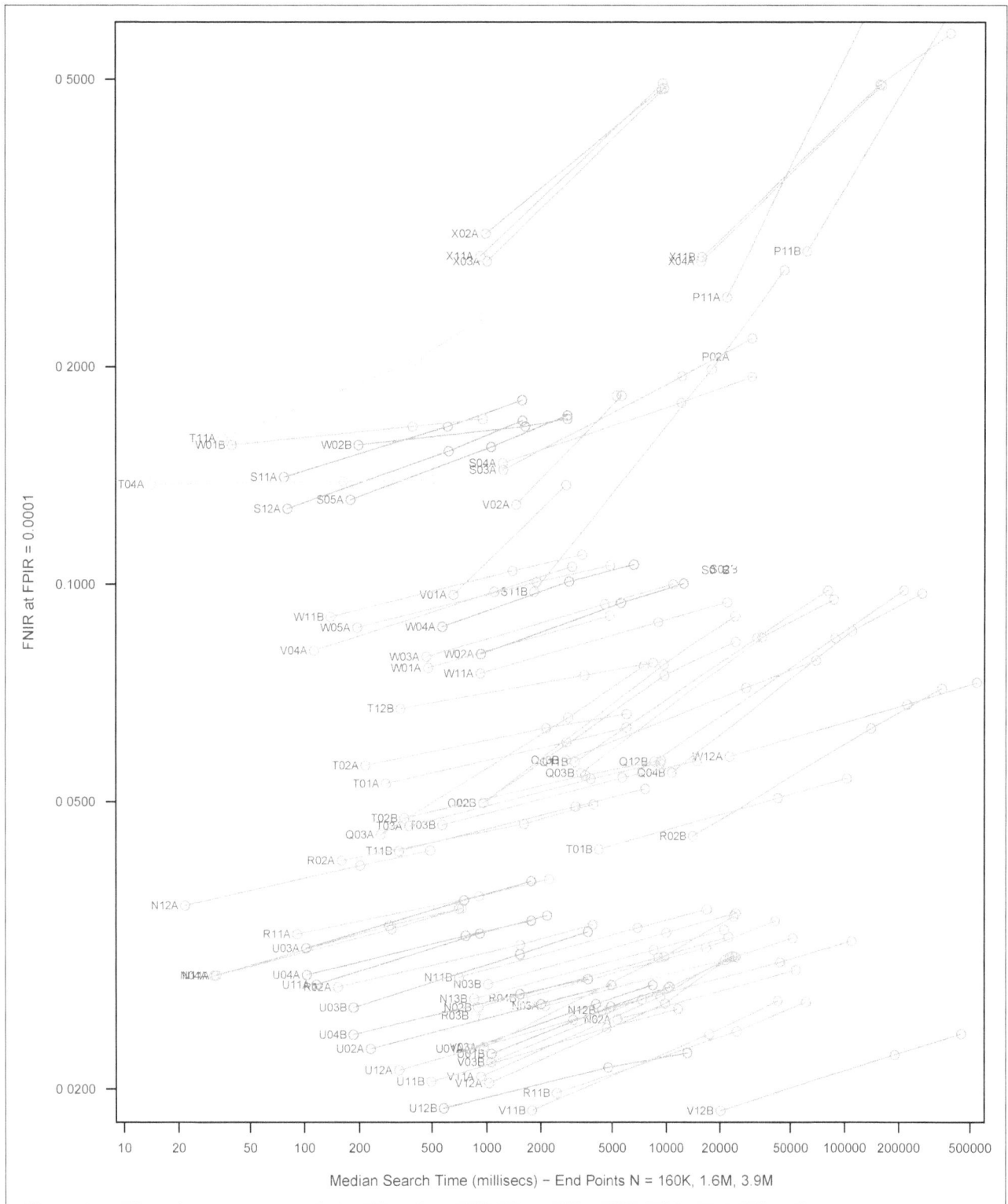

Figure 12: **Scalability tradespace**: Each line connects three points corresponding to populations N = 160K, N = 1,6M, and N = 3.9M. The x-axis plots median template search duration. The y-axis plots FNIR at FPIR = 0.0001. Short horizontal extent shows better-than-linear scaling of search duration. Large vertical extent reveals poor recognition accuracy. The color hue is linear in $\log_{10}(T_2/T_1)$ corresponding to Power-law search speed dependency. The exact values can be computed from the abscissae or from Tables 9 and 10. Darker colors correspond to better-than-linear scalability. Times are estimated on the IREX API-specified NIST blade. Accuracy is estimated executing searches from the large search set, $\mathcal{S}_{1b}$ in enrolled populations of single eyes. The figure excludes 330x330 images. Single points appear because one timing or accuracy estimate could not be completed.

| FNIR = FALSE NEGATIVE IDENT. RATE | N = NEUROTECHNOLOGY | P = SMU | Q = IRITECH | R = COGENT | S = SMARTSENSORS | T = CAMBRIDGE |
| FPIR = FALSE POSITIVE IDENT. RATE | U = L1 | V = MORPHO | W = IRISID | X = CROSSMATCH | Y = KYNEN | |

5   Speed-accuracy tradeoffs

Both accuracy and speed are important to most applications of biometric identification technology. In the general case these quantities can be traded off against each other. For example, in fingerprint identification, AFIS implementations often apply different algorithms to different filtered partitions of the entire population, the heavyweight algorithms being applied only to the end-stage likely candidates. In IREX III, NIST solicited submission of fast vs. slow and experimental vs. mature variants with the explicit statement that IREX would report speed-accuracy tradeoff figures of the form shown in Figure 13. This exposes algorithms that are accurate but computationally expensive.

5.1   Methods

Figures 13 to 15 plot an accuracy number (FNIR at FPIR = 0.0001 for N = 1.6 million) for a particular implementation against the durations of its template generation and search functions and the combined total of those two. The figures additionally color-code each data point by either template-size or the template generation time. The figures plot data that are tabulated elsewhere in this document.

5.2   Results

Template Generation: Regarding the accuracy vs. template generation cost tradespace, Figure 13 shows

▷ Large templates, shaded in lighter colors, are typically produced more slowly. This implies that more elaborate feature encodings are more costly to compute. Template sizes are tabulated in Tables 6 and 7. Template generation times, which span a factor of almost 30, are presented in Table 8.

▷ For many providers, variations in template generation times are small and variations are not strongly correlated to accuracy changes (see for example R, T, W, X implementations). However, other providers, (e.g. Q, S, U and V), have separated clusters; these, together with template sizes indicate that different underlying feature representations are being used. Thus, for example, SxxA and SxxB, differ in speed by a factor of almost 10 (70 vs. 700 msec), and QxxA and QxxB durations span times from 30 milliseconds to nearly 500.

▷ The most accurate implementation, V12B, uses one of the most computationally expensive template generators. However, almost identical accuracy can be achieved using the U12B algorithm that runs about four times faster: 660 vs. 160 milliseconds.

Search: Regarding the accuracy vs. search cost tradespace, Figure 14 shows that:

▷ For most providers, search duration varies more widely between implementations than that for template generation. Search cost is also more related to accuracy variations, although the association is not particularly strong in either case. For example, the search speeds of the N algorithms vary by at least a factor of 200 (N11A and N12A vs. N02A and N12B) with little improvement in accuracy. Similarly R12A and R11B are a factor of ten apart in speed, but close in accuracy.

▷ There are cases where additional cost gives an accuracy benefit. While V12B is 200 times slower than V04A and gives almost 5 times fewer false rejections, a middle ground is available using V12A which is ten times slower with more than 3 times fewer errors. However, the tradeoff is one of diminishing returns: R11B produces about 20% fewer false rejections than R12A, yet is at least ten times more slowly.

▷ In some cases, the more costly algorithms are less accurate. For example, R02B is the slowest and least accurate R implementation. Likewise, S01B and S02B are no more accurate than the S1xA siblings that run 300 times faster. The most accurate T algorithm, T11B, is more accurate and at least ten times faster than T01B. The position is similar with X03A and X04A.

FNIR = FALSE NEGATIVE IDENT. RATE	N = NEUROTECHNOLOGY	P = SMU	Q = IRITECH	R = COGENT	S = SMARTSENSORS	T = CAMBRIDGE
FPIR = FALSE POSITIVE IDENT. RATE	U = L1	V = MORPHO	W = IRISID	X = CROSSMATCH	Y = KYNEN	

5.3 Conclusion

The conclusion of this section is that many providers can trade accuracy for speed, but not so much as to attain the speed of the fastest providers, nor the low FNIR of the most accurate. The fact that speed and accuracy vary widely and not along a clearly defined frontier is a reflection that some providers' algorithms are more capable than others, and variation is due to more than just refinement of parameters. This is partially attributable to participants having variable access to meaningful development data. Thus, while implementers were given limited performance data over the course the IREX III project, they were not given quantified cause-and-effect data regarding algorithm failure and so could not make specific algorithmic modifications to reduce FNIR. Note that algorithm providers were repeatedly made aware that both accuracy and speed would be measured and reported.

End users of these algorithms should select specific implementations that satisfy their operational requirements. It is not known, however, whether the providers have instituted sufficient version controls to respond to specific algorithm requests. Appendix J of this report includes checksums of the libraries provided to IREX III; these can be used by implementers to trace which algorithms and parameters were used to achieve the speed and accuracy measurements reported here. End users are cautioned that selection of a provider alone is usually insufficient to ensure operational constraints can be met.

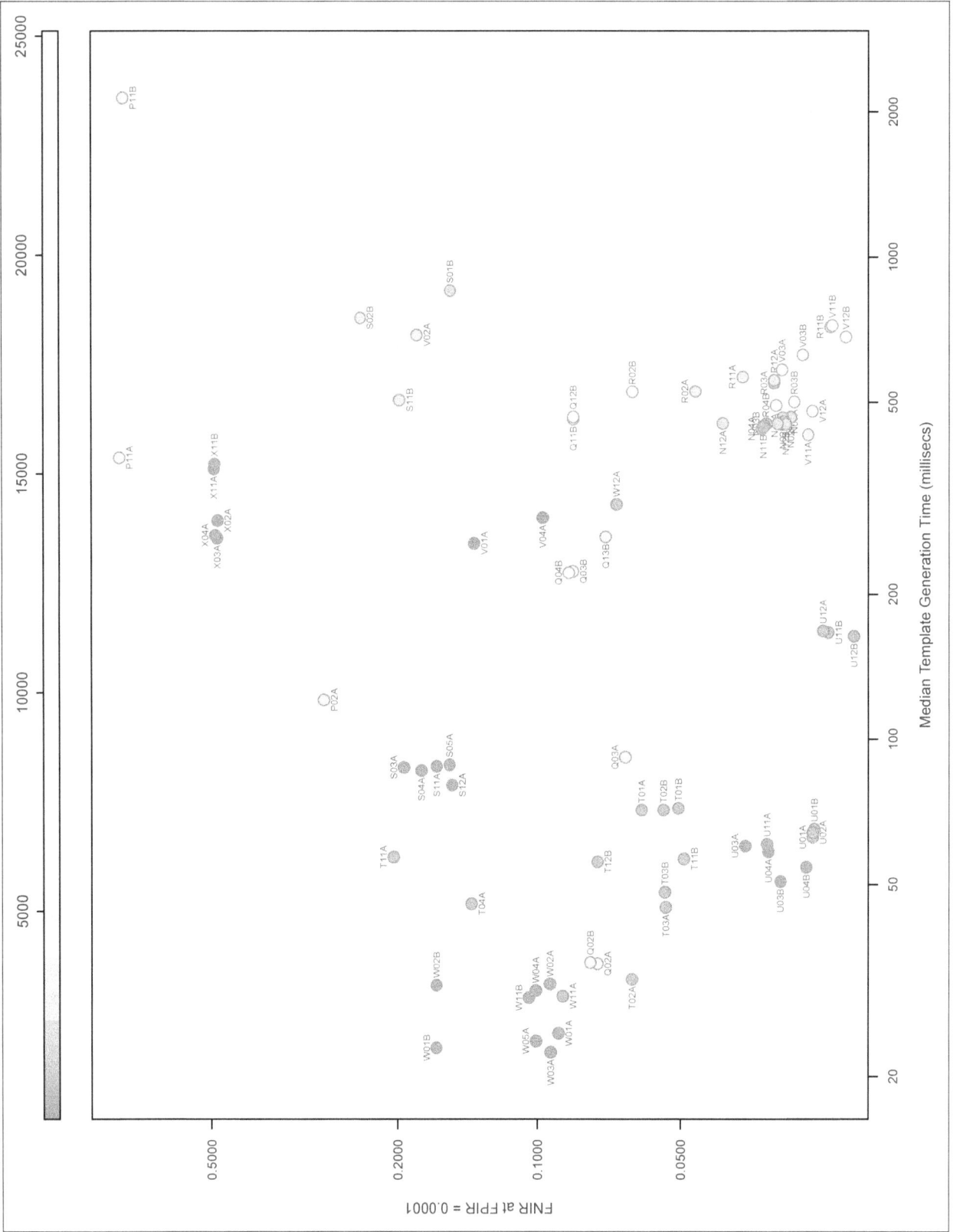

Figure 13: **Template generation speed vs. accuracy tradespace:** The plots show the miss rate (FNIR at FPIR = 0.0001) against the median duration of the template generation call. The timing estimates apply to the IREX III API-specified LINUX blade, producing templates from the large search set, S_{1b}. The figure excludes 330x330 images; this is more representative of future applications. The points are colored according to the size of the enrollment templates, in bytes.

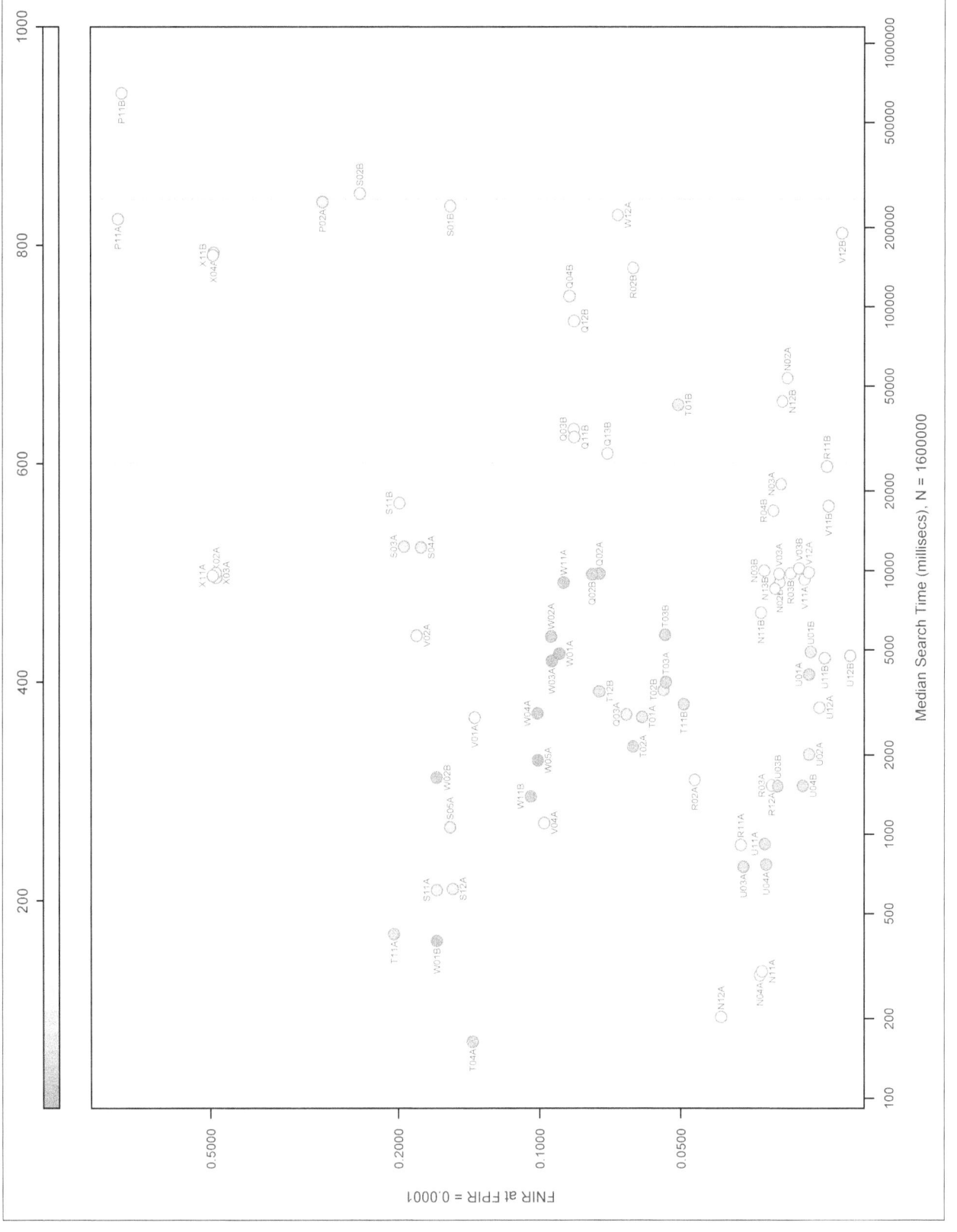

Figure 14: **Search speed vs. accuracy tradespace**: The plots show the miss rate (FNIR at FPIR = 0.0001) against the median duration of the template search function. The timing estimates apply to the IREX III API-specified LINUX blade, executing searches from the large search set, S_{1b}, in an enrolled population of N = 1, 600, 000 single eyes. The points are colored according to the duration of template generation function. The figure excludes 330x330 images; this is more representative of future applications. The vertical lines indicate the speed limits for class A and B submission established in the IREX III API, CONOPS AND EVALUATION PLAN document

| FNIR = FALSE NEGATIVE IDENT. RATE | N = NEUROTECHNOLOGY | P = SMU | Q = IRITECH | R = COGENT | S = SMARTSENSORS | T = CAMBRIDGE |
| FPIR = FALSE POSITIVE IDENT. RATE | U = L1 | V = MORPHO | W = IRISID | X = CROSSMATCH | Y = KYNEN | |

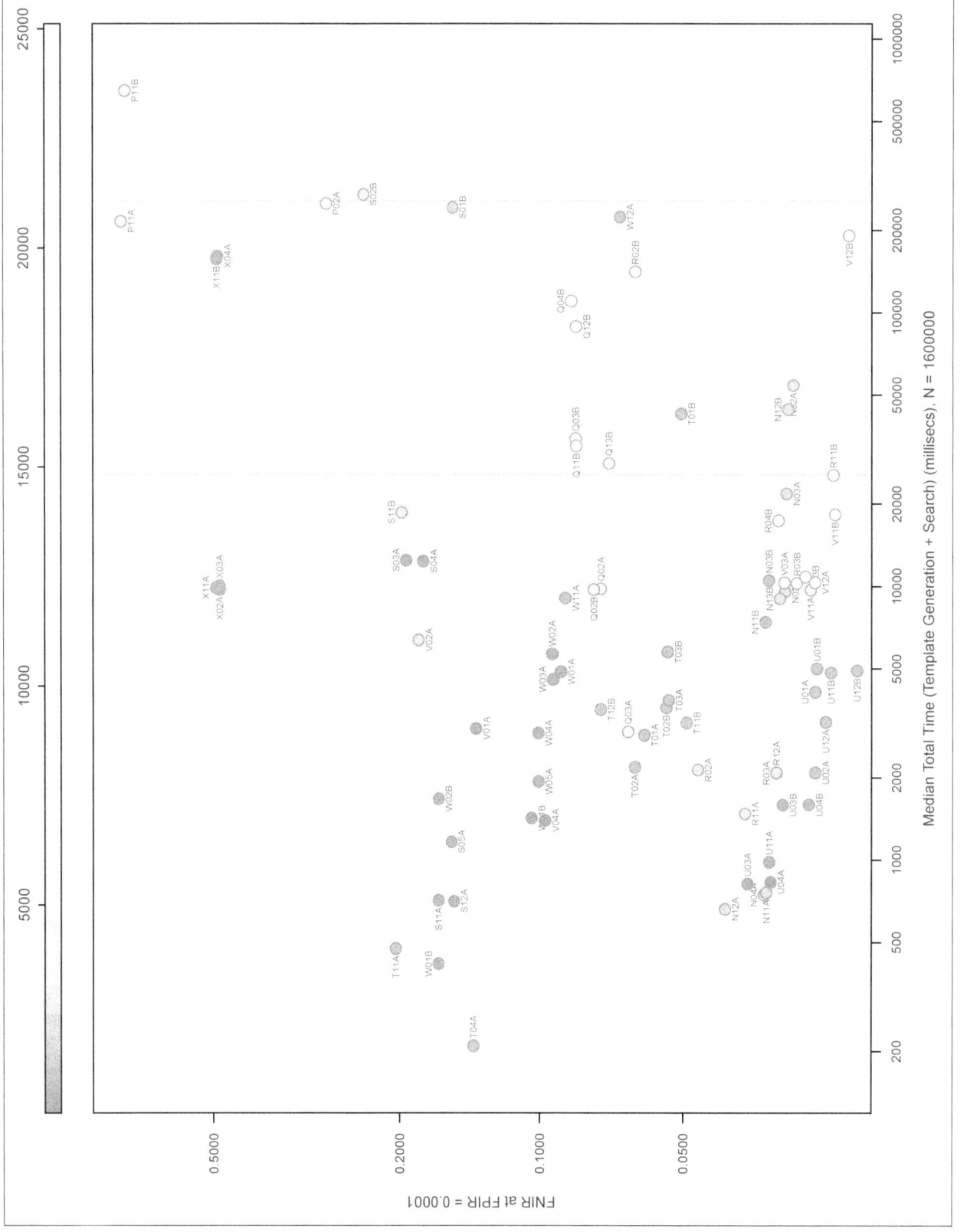

Figure 15: **Combined image-search speed vs. accuracy tradespace:** The plots show the miss rate (FNIR at FPIR = 0.0001) against the median of the sum of the template generation and template search durations This represents the end-to-end search time, using the IREX III API-specified NIST blade, executing searches from the large search set, S_{1b}, in an enrolled population of N = 1, 600, 000 single eyes The points are colored according to the size of the enrollment templates, in bytes The figure excludes 330x330 images; this is more representative of future applications The vertical lines indicate the speed limits for class A and B submission established in the IREX III API, CONOPS AND EVALUATION PLAN document

| FNIR = FALSE NEGATIVE IDENT. RATE | N = NEUROTECHNOLOGY | P = SMU | Q = IRITECH | R = COGENT | S = SMARTSENSORS | T = CAMBRIDGE |
| FPIR = FALSE POSITIVE IDENT. RATE | U = L1 | V = MORPHO | W = IRISID | X = CROSSMATCH | Y = KYNEN | |

6 Threshold-based accuracy

6.1 Rationale

The majority of biometric identification applications include the application of a decision threshold. This threshold implements a policy on the tradeoff between false negative and false positive outcomes. A threshold might be applied during the search as part of the algorithms internals[14] and also post-search, by the operator, or in this case, the test laboratory. In most applications, a low threshold is adopted and a candidate list will have length one or zero: One will indicate a high likelihood that the search sample and the hypothesized mate are indeed from the same person. Zero will indicate that the person submitting the search sample has no prior enrolled mate, *or* that the search sample is defective (e.g. blur, or closed eyes) and a miss has occurred. Either way, when used in this *identification* mode, the iris recognition system must be configured with an operating threshold.

6.2 Metrics

The test protocol requires that, for every search, the implementation under test reports an ordered list of hypothesized matching candidates. Unless stated otherwise, the implementation was asked to produce $L = 20$ candidates. The experimental design is to enroll exactly one sample[15] for each person, and then to execute two kinds of searches. First is a *mate* search for which exactly one candidate should have a low dissimilarity value, and for which $L-1$ candidates should have higher dissimilarity values. That is the ideal outcome. In practice, poor quality images sometimes produce matching entries at high dissimilarity values. More frequently, the correct enrolled candidate is not present in the top L candidates. The second kind of search is termed a *nonmate* search, for which all L candidates should produce high dissimilarity values, because there are zero enrolled elements for this person.

▷ **FPIR**: A false positive occurs when a person not enrolled in the system is matched against someone else who is. This is an undesirable outcome: In negative identification systems (e.g. benefits fraud, criminal detection) it implicates or derogates someone. In positive identification systems (e.g. one-to-many access control) it incorrectly extends benefit to someone. The relevant metric is the fraction of nonmate searches that produce nonmates below threshold τ on a candidate list of length L. This is referred to as the false positive identitification rate, $\mathrm{FPIR}(\tau)$. It is estimated over Q searches for which there is no enrolled mate, and is defined here as

$$\mathrm{FPIR}(\tau) = \frac{\sum_{q=1}^{Q}\sum_{r=1}^{L} 1 - H(d_{qr} - \tau)}{\sum_{q=1}^{Q}\sum_{r=1}^{L} 1 - H(d_{qr} - \infty)} = \frac{\sum_{q=1}^{Q}\sum_{r=1}^{L} 1 - H(d_{qr} - \tau)}{QL} \qquad (1)$$

where L is the length of the candidate list, here $L = 20$, and d_{qr} is the r-th lowest dissimilarity reported by the algorithm for the q-th search. The function $H(x)$ is the Heaviside step function

$$H(x) = \begin{cases} 0 & x < 0 \\ 1 & x \geq 0 \end{cases} \qquad (2)$$

▷ **FNIR**: A false negative occurs when a person enrolled in a system is incorrectly not matched against other samples of the same person. This is an undesirable outcome: In negative identification systems (e.g. benefits fraud, criminal detection) the sought person proceeds undetected. In positive identification systems (e.g. one-to-many access control) it requires the user to make further attempts to match, or proceed to some secondary process. The relevant metric is the fraction of searches for which the enrolled mate is not returned below a threshold τ on a candididate list of length L. This is referred to as the false negative identification rate, $\mathrm{FNIR}(\tau)$. It is estimated over P searches

---

[14]It is well known that some implementations do not report scores above a certain threshold, (e.g. a Hamming Distance greater than 0.33) because efficiency gains can be realized by only computing partial template comparisons, short circuiting the complete distance calculation above some level.

[15]A sample here contains one or more iris images, from one or more eyes, bundled into a MULTIIRIS data structure - see Figure 22 for a two-eye example.

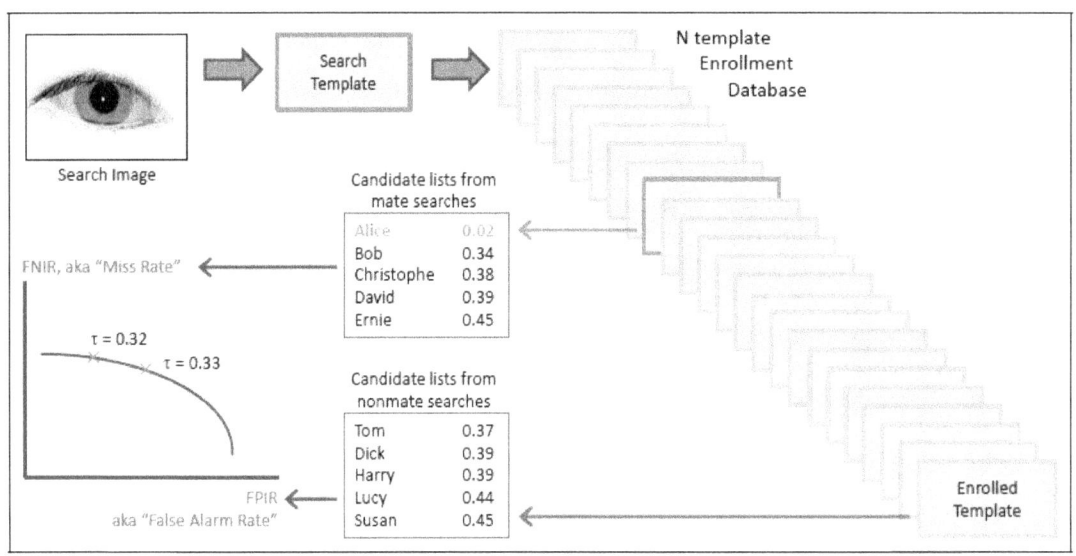

Figure 16: Schematic of the one-to-many testing protocol. The two error rates are computed over many searches.

for which there is an enrolled mate, and is defined formally as

$$\text{FNIR}(\tau) = 1 - \frac{1}{P} \sum_{p=1}^{P} \sum_{r=1}^{L} I_{pr} \left[1 - H(d_{pr} - \tau) \right] \tag{3}$$

where d_{pr} is the r-th lowest dissimilarity reported by the algorithm for the p-th search, and I_{pr} is 1 only if the identity of the r-th candidate is the same as the identity of search p, and 0 otherwise. In the few cases where algorithms placed the mate on a candidate list twice, it was ignored. If the algorithm placed the correct mate on the candidate list more than once, only the lowest (best) rank for the mate was counted.

6.3   Effect of ground truth errors

The operational database contains ground-truth identity errors and these have effects on the FNIR and FPIR accuracy estimates. There are two types of ground truth error:

▷ **Type I:** One person's images are present under two or more identities In testing, these occurrences lead to apparent false positives since images the laboratory believes should not match, actually do. This is the cause of the unexpected shape of the red DET in Figure 17. The curve rises sharply because this nonmate distribution is contaminated with low-dissimilarity same-eye comparisons. Rather than let the problem persist, NIST elected to estimate FPIR using search images that have been flipped[16] about a vertical axis i.e. L-R mirroring such that left eyes appear as right eyes. The effect of this is to manufacture nonmate comparisons by relying on the fact that iris recognition algorithms are not invariant under reflection (a mirror image of an image will not match the original). This removes the effect of Type I ground truth errors[17]. This legerdemain was presented to all IREX III participants; none objected. It is

---

[16] This was implemented using the lossless *jpegtran* application provided by the Independent JPEG Group, and present on most LINUX platforms.

[17] The mirroring is effective for the fraction of irides, p, for which a mate truly does not exist under a different ID. For the small problematic fraction, $1 - p$, where a mate is erroneously present, i.e. the fraction responsible for the aberrant DET shape, a qualification is necessary. These comparisons are of an iris with the mirror of another image of itself. The mirroring means that the iris comparison looks like a nonmate comparison only to the extent that an iris' texture is not self-similar under reflection. This is mostly true, but at least two effects imbue self-similarity: First the same eye may be affected by persistent effects such as ptosis and pupillary constriction; second it is known that spatial correlation exists in features extracted from iris images, but it is not known that such correlation exists under mirror reflection. In summary, the mirror imaging is potentially less effective for the small fraction $1 - p$ of irides that are present under two or more IDs. In any case, this will still lead to an overestimation of FPIR, i.e. a conservative outcome.

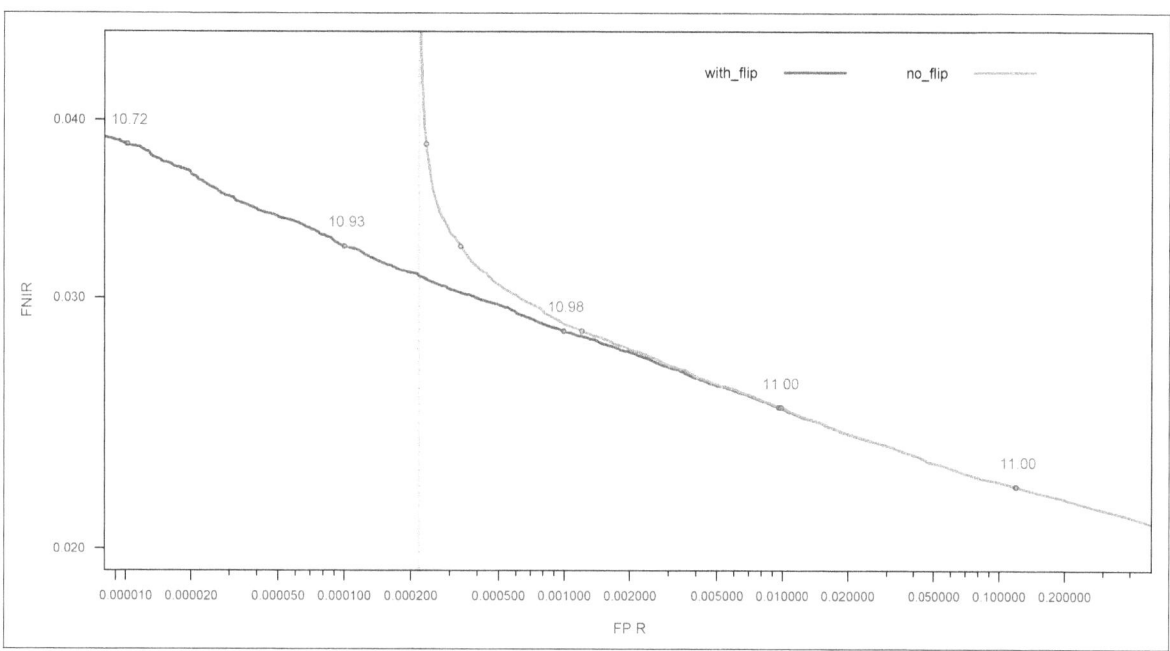

Figure 17: For the U04A SDKs two identification DETs are presented. The red line shows the DET when FPIR is computed with the images as provided from the original operational source. The blue line shows the DET when FPIR is computed from nonmate searches of left-right mirror image flipped images against unflipped enrollment images. The green line is added to show asymptotic behavior. The mirroring is implemented to avoid effects of ground truth errors - See the discussion of section 6.3. The population size is N = 1,600,000.

effective in reducing FPIR - the blue DET in Figure 17, and, while it was applied to avert adverse reporting of iris false matching performance, the issue of ground truth errors in operational databases is real and important. Particularly in a deployment, a search will usually hit all the enrolled identities unless the cause of the problem is quality-related. The operator should detect the matches and take appropriate steps to consolidate the records under one identity. Thus, the elevated workload implications on backend system operators (if the occurrences are recognized) can be shifted to the front-end image collection workforce - where greater training, care, and process refinements will ameliorate the introduction of ground truth bugs. This will likely have cost implications.

▷ **Type II**: Images of two or more persons share the same identity. In testing, these cases lead to apparent false negatives since images the laboratory believe should match, actually do not. Elevated FNIR affects the IREX III results. The exact extent of the problem is not known, however. The best observed rank-one miss rate of around 1.5% (V11B, Table 12) is axiomatically an upper bound, and the results given in the IREX III FAILURE ANALYSIS SUPPLEMENT suggest ground-truth errors are responsible for only a small fraction of measured FNIR values. In real operations, a new search will usually succeed but the operator who reviews all the images will need to notice the incorrect identity merging and split the record (after appropriate duplicate detection and visual review).

6.4 Treatment of candidate lists

In some cases the implementation did not produce candidate lists, or produced short candidate lists.

▷ **FTX and FTS**: In some cases the implementation failed to produce a template from the search imagery. This is termed *failure to extract* (FTX). In other cases the implementation produced a template but failed to return any candidates. This is termed *failure to search* (FTS). FTX and FTS are treated as failures in a positive identification system.

(a) Group 0

	Single Eye		Two eyes	
N IDs	1600000	553230	1600000	315662
N Images	1600000	553230	3123677	617473
	Enrol	Search	Enrol	Search
N02A	0.0002	0.0002	0.0000	-
N02B	0.0002	0.0002	0.0000	0.0000
N03A	0.0002	0.0002	0.0000	0.0000
N03B	0.0002	0.0002	0.0000	0.0000
N04A	0.0002	0.0002	0.0000	0.0000
P02A	0.0664	-	0.0376	-
Q02A	0.0000	0.0000	0.0000	0.0000
Q02B	0.0000	0.0000	0.0000	0.0000
Q03A	0.0000	0.0000	0.0000	0.0000
Q03B	0.0000	0.0000	0.0000	0.0000
Q04B	0.0000	0.0000	0.0000	0.0000
R02A	0.0000	0.0000	0.0000	0.0000
R02B	0.0000	0.0000	0.0000	-
R03A	0.0000	0.0002	0.0000	0.0001
R03B	0.0000	0.0000	0.0000	0.0000
R04B	0.0000	0.0000	0.0000	0.0000
S01B	0.0002	-	0.0002	-
S02B	0.0001	-	0.0001	-
S03A	0.0100	0.0076	0.0037	0.0032
S04A	0.0000	0.0000	0.0000	0.0000
S05A	0.0168	0.0163	0.0186	0.0153
T01A	0.0000	0.0000	0.0000	0.0000
T01B	0.0000	0.0000	0.0000	0.0000
T02A	0.0000	0.0000	0.0000	0.0000
T02B	0.0000	0.0000	0.0000	0.0000
T03A	0.0000	0.0000	0.0000	0.0000
T03B	0.0000	0.0000	0.0000	0.0000
T04A	0.0000	0.0000	0.0000	0.0000
U01A	0.0001	0.0001	0.0000	0.0000
U01B	0.0001	0.0001	0.0000	0.0000
U02A	0.0001	0.0001	0.0000	0.0000
U03A	0.0000	0.0000	0.0000	0.0000
U03B	0.0000	0.0000	0.0000	0.0000
U04A	0.0000	0.0000	0.0000	0.0000
U04B	0.0000	0.0000	0.0000	0.0000
V01A	0.0000	0.0000	0.0000	0.0000
V02A	0.0000	0.0000	0.0000	0.0000
V03A	0.0000	0.0000	0.0000	0.0000
V03B	0.0000	0.0000	0.0000	0.0000
V04A	0.0000	0.0000	0.0000	0.0000
W01A	0.0011	0.0010	0.0020	0.0019
W01B	0.0000	0.0000	0.0000	0.0000
W02A	0.0000	0.0000	0.0000	0.0000
W02B	0.0000	0.0000	0.0000	0.0000
W03A	0.0000	0.0000	0.0000	0.0000
W04A	0.0000	0.0000	0.0000	0.0000
W05A	0.0000	0.0000	0.0000	0.0000
X02A	0.0376	0.0336	-	-
X03A	0.0075	0.0067	-	-
X04A	0.0075	0.0067	-	-
Y02A	-	-	-	-
Y02B	-	-	-	-
Y03A	0.0121	-	-	-
Y03B	0.0017	-	-	-

(b) Group 1

	Single Eye		Two eyes	
N IDs	1600000	553230	1600000	315662
N Images	1600000	553230	3123677	617473
	Enrol	Search	Enrol	Search
N11A	0.0002	0.0002	0.0000	0.0000
N11B	0.0002	0.0002	0.0000	0.0000
N12A	0.0002	0.0002	0.0000	0.0000
N12B	0.0046	0.0049	0.0002	0.0002
N13B	0.0046	0.0049	0.0002	0.0002
P11A	0.0664	0.0656	0.0376	-
P11B	0.0000	0.0000	0.0000	-
Q11B	0.0000	0.0000	0.0000	0.0000
Q12B	0.0000	0.0000	0.0000	0.0000
Q13B	0.0000	0.0000	0.0000	0.0000
R11A	0.0000	0.0002	0.0000	0.0001
R11B	0.0000	0.0000	0.0000	0.0000
R12A	0.0000	0.0002	0.0000	0.0001
S11A	0.0000	0.0000	0.0000	0.0000
S11B	0.0028	0.0024	0.0026	-
S12A	0.0000	0.0000	0.0000	0.0000
T11A	0.0000	0.0000	0.0000	0.0000
T11B	0.0000	0.0000	0.0000	0.0000
T12B	0.0000	0.0000	0.0000	0.0000
U11A	0.0001	0.0000	0.0000	0.0000
U11B	0.0001	0.0000	0.0000	0.0000
U12A	0.0001	0.0000	0.0000	0.0000
U12B	0.0001	0.0000	0.0000	0.0000
V11A	0.0000	0.0000	0.0000	0.0000
V11B	0.0000	0.0000	0.0000	0.0000
V12A	0.0000	0.0000	0.0000	0.0000
V12B	0.0000	0.0000	0.0000	0.0000
W11A	0.0000	0.0000	0.0000	0.0000
W11B	0.0000	0.0000	0.0000	0.0000
W12A	0.0000	0.0000	0.0000	0.0000
X11A	0.0072	0.0065	0.0017	0.0015
X11B	0.0072	0.0065	0.0017	-

Table 11: The fraction of images that did not produce a template during enrollment and search. Template generation failure is detected if the function call returned a non-zero error code, threw an exception or crashed, or executed for longer than twenty seconds (i.e. forty times the IREX III 90-th percentile specification). Green coloring indicates that the software never produced a failure. The group designations 0 and 1 indicate SDK submission in February-June and Aug 2011, respectively.

FNIR = FALSE NEGATIVE IDENT. RATE	N = NEUROTECHNOLOGY	P = SMU	Q = IRITECH	R = COGENT	S = SMARTSENSORS	T = CAMBRIDGE
FPIR = FALSE POSITIVE IDENT. RATE	U = L1	V = MORPHO	W = IRISID	X = CROSSMATCH	Y = KYNEN	

- – Nonmate searches: The L dissimilarity values of the candidate list are set to a very high value, guaranteeing correct rejection.

- – Mate searches: The L dissimilarity values of the candidate list are set to a very high value, guaranteeing a false rejection.

This combination is only punitive for genuine searches - the nonmate searches give the correct result, rejection. This is inappropriate because the outcome is obtained for the wrong reason (FTX, rather than biometric rejection). The fractions are are shown in Table 11.

▷ **Partial candidate lists**: In some cases the implementation returns $R_D < R$ candidates, and $R - R_D$ null entries. Usually R_D is a random-variable. This might arise as an intentional efficiency strategy or as a result of a property of the search algorithm. While this is not an adverse outcome it requires proper handling, as follows. If the hypothesized identity is not a valid entry in the enrollment database, any reported dissimilarity value is replaced by a very high value. The effect of this depends on the kind of search:

- – Nonmate searches: The high values guarantee correct rejection.

- – Mate searches: The high values guarantee a false negative outcome (a miss), unless the correct mate appears elsewhere on the candidate list.

6.5 Results

This section summarizes identification accuracy via tables, graphs and discussion. In addition this report's accompanying appendices include algorithm-specific performance statements. These figures and tables, produced automatically for each of the 85 implementations tested, amount to several hundred pages and are too numerous to include here. The results are of primary interest to the recognition algorithm developers and engineers considering deployment. In the various figures, results for some algorithms are missing or marked only with a hyphen. These occurences arise because the run was not attempted or not completed successfully, typically due to crashes, hangs, or timeouts, or inability to handle large populations.

The results of the main body of the report are as follows.

▷ Table 12 shows one-eye FNIR values for five fixed FPIR values. These are estimated for searches of sets $\mathcal{S}_{1b}$ into an enrolled population of $N = 1,600,000$.

▷ Table 13 shows two-eye results estimated over the small ($\mathcal{S}_{2a}$) and large ($\mathcal{S}_{2b}$) search sets.

▷ Figures 18 and 19 show one-eye DET characteristics for all algorithms for N = 1,600,000 estimated over the large set $\mathcal{S}_{1b}$.

▷ Figures 20 and 21 show one-eye DET characteristics for selected algorithms for N = 3,904,239 estimated over the large set $\mathcal{S}_{1b}$.

▷ Figures 23 and 24 show two-eye DET characteristics for all algorithms for N = 1,600,000 estimated over the large set $\mathcal{S}_{2b}$.

▷ Figures in Appendix H in the IREX III APPENDICES[18] show one- and two-eye DETs for 640x480, 480x480 and 330x330 images and the general population.

The notable observations are as follows:

---

[18]Linked from http://iris.nist.gov/irex

▷ **Range**: False negative identification error rates (i.e. miss rates) vary by an order of magnitude. For single-eye, the most accurate implementation misses just 1.8% of mates (V03B, FNIR = 0.018 at FPIR = 0.001, Table 12 excluding 330x330). The least accurate implementations miss 20% of mates. To some unknown extent these error rates would be improved if the image quality was increased via improved design of cameras, collection practices and database integrity controls.

▷ **Value of setting a threshold**: Many of the DET curves have a low enough gradient that operating the iris algorithms at a high (i.e. weak, non-discriminating) threshold produces only marginally more mates. For the leading algorithms, when FPIR is allowed to vary between 0.1 and 0.0001 such that false matches are 1000 times less frequent, the FNIR values increase by around 50% (e.g. U04B, V03A, R03A, N03A, Table 12).

▷ **Possible ground-truth errors**: Rank one miss rates are sometimes higher than FNIR at FPIR = 0.001 (e.g. U04A, Tables 12 a and b). This occurs because some similar mates are being found at other than rank 1. This may be due to the presence of ground-truth errors in the database, particularly when an eye under the wrong ID is ranked better than an eye with the correct ID.

▷ **Pathological 330x330 images**: The highly compressed 330x330 images contribute heavily to miss rates. The most accurate implementations find at least five times fewer mates with these images than with other images (640x480, 480x480) (see Figure 4 and Appendix H).

▷ **Evolution of performance**: Over the 8 month span of the testing activity, and the 12 month span of the entire IREX III evaluation, some providers improved accuracy (R, V, and, somewhat, U and S) while others essentially did not (Q, N, W, X, T, P). This is probably a negative result for IREX III which included progressive improvement in its aims. That said, while search speed did improve in some cases (T, U), this was accompanied by a net slowing down of algorithms likely in pursuit of improved accuracy.

▷ **Shape of the DET characteristic**: Many of the DET curves have an approximately fixed gradient across the range $10^{-5} \leq \text{FPIR} \leq 10^{-1}$. This is true despite the vertical FNIR offset of the lines varying considerably (e.g. R03A vs. W04A). Other DETs curve upwards as FPIR decreases. These effects are related to the consistency and accuracy of the localization of iris texture, handling of rotation, and the innate strength of the feature representation of the iris.

| FNIR = FALSE NEGATIVE IDENT. RATE | N = NEUROTECHNOLOGY | P = SMU | Q = IRITECH | R = COGENT | S = SMARTSENSORS | T = CAMBRIDGE |
| FPIR = FALSE POSITIVE IDENT. RATE | U = L1 | V = MORPHO | W = IRISID | X = CROSSMATCH | Y = KYNEN | |

Table 12: FNIR at various FPIR values, for single-eye enrollment and search, estimated using set S_{1b}. The size of enrolled population is $N = 1,600,000$. The poor quality 330x330 images are not included. The Table corresponds to Figures 18 and 19. The narrow sub-tables give rank one miss rates (β values from equation 13). Cells are shaded light green when FNIR ≤ 0.05, and dark green for FNIR ≤ 0.025. A row containing hyphens indicates the large-set run did not complete successfully, typically due to an ability to handle large N, crashes, hangs, or timeouts. In these cases, β values are stated in blue for the small-set S_{2b}, if available. Two-eye results appear in Table 13.

(a) FNIR, Group 0   (b) Rank 1, Group 0

SDK	0.00001	0.0001	0.001	0.01	0.1	β(1)	β(20)
N02A	0.036	0.029	0.025	0.022	0.019	0.018	0.015
N02B	0.039	0.030	0.026	0.023	0.020	0.019	0.015
N03A	0.038	0.030	0.026	0.022	0.019	0.018	0.015
N03B	0.040	0.033	0.028	0.024	0.020	0.020	0.016
N04A	0.040	0.034	0.029	0.025	0.022	0.022	0.018
P02A	-	-	-	-	-	0.161	0.138
Q02A	0.112	0.075	0.045	0.032	0.027	0.028	0.025
Q02B	0.128	0.077	0.045	0.031	0.026	0.027	0.023
Q03A	0.120	0.065	0.037	0.030	0.027	0.029	0.027
Q03B	0.135	0.084	0.037	0.024	0.021	0.023	0.021
Q04B	0.138	0.086	0.039	0.023	0.019	0.022	0.019
R02A	0.053	0.047	0.041	0.037	0.034	0.034	0.032
R02B	0.085	0.063	0.046	0.031	0.025	0.027	0.023
R03A	0.038	0.032	0.032	0.025	0.023	0.025	0.022
R03B	0.034	0.029	0.025	0.023	0.021	0.022	0.019
R04B	0.039	0.031	0.027	0.024	0.022	0.022	0.020
S01B	-	-	-	-	-	-	-
S02B	-	-	-	-	-	0.054	0.039
S03A	0.284	0.194	0.141	0.107	0.084	0.090	0.067
S04A	0.220	0.178	0.145	0.119	0.099	0.104	0.083
S05A	0.202	0.155	0.127	0.108	0.094	0.096	0.081
T01A	0.068	0.060	0.053	0.047	0.042	0.043	0.037
T01B	0.059	0.051	0.043	0.037	0.031	0.031	0.025
T02A	0.071	0.063	0.056	0.051	0.046	0.047	0.042
T02B	0.061	0.054	0.048	0.042	0.038	0.038	0.033
T03A	0.061	0.054	0.047	0.040	0.035	0.036	0.030
T03B	0.062	0.054	0.047	0.041	0.036	0.036	0.030
T04A	0.141	0.139	0.138	0.137	0.136	0.137	0.136
U01A	0.031	0.026	0.023	0.019	0.017	0.018	0.014
U01B	0.030	0.026	0.022	0.019	0.016	0.017	0.016
U02A	0.031	0.026	0.023	0.019	0.017	0.018	0.014
U03A	0.044	0.037	0.031	0.027	0.023	0.024	0.020
U03B	0.037	0.031	0.026	0.022	0.019	0.020	0.016
U04A	0.038	0.033	0.028	0.025	0.022	0.024	0.020
U04B	0.031	0.027	0.024	0.021	0.018	0.020	0.016
V01A	0.299	0.137	0.092	0.075	0.061	0.066	0.050
V02A	0.281	0.182	0.097	0.043	0.035	0.039	0.032
V03A	0.049	0.031	0.022	0.019	0.016	0.017	0.015
V03B	0.048	0.028	0.021	0.018	0.016	0.017	0.015
V04A	0.121	0.098	0.080	0.067	0.055	0.059	0.048
W01A	0.105	0.090	0.077	0.066	0.056	0.057	0.052
W01B	0.176	0.165	0.156	0.148	0.142	0.143	0.137
W02A	0.109	0.094	0.080	0.068	0.056	0.056	0.042
W02B	0.176	0.165	0.156	0.148	0.142	0.143	0.137
W03A	0.109	0.094	0.080	0.067	0.055	0.055	0.042
W04A	0.115	0.101	0.088	0.076	0.065	0.065	0.052
W05A	0.115	0.101	0.087	0.075	0.064	0.065	0.052
X02A	0.801	0.485	0.286	0.179	0.118	0.128	0.096
X03A	0.805	0.486	0.265	0.148	0.085	0.093	0.064
X04A	0.805	0.491	0.265	0.147	0.081	0.090	0.059
Y02A	-	-	-	-	-	-	-
Y02B	-	-	-	-	-	-	-
Y03A	-	-	-	-	-	-	-
Y03B	-	-	-	-	-	-	-

(c) FNIR, Group 1   (d) Rank 1, Group 1

SDK	0.00001	0.0001	0.001	0.01	0.1	β(1)	β(20)
N11A	0.041	0.033	0.029	0.025	0.022	0.021	0.018
N11B	0.041	0.033	0.029	0.024	0.021	0.020	0.017
N12A	0.048	0.041	0.036	0.033	0.030	0.029	0.026
N12B	0.037	0.030	0.026	0.023	0.020	0.019	0.016
N13B	0.040	0.031	0.027	0.023	0.020	0.019	0.016
P11A	0.964	0.785	0.400	0.208	0.164	0.193	0.142
P11B	0.946	0.771	0.427	0.221	0.150	0.177	0.112
Q11B	0.143	0.084	0.056	0.039	0.033	0.034	0.028
Q12B	0.143	0.082	0.055	0.037	0.031	0.031	0.025
Q13B	0.105	0.072	0.056	0.047	0.039	0.039	0.032
R11A	0.043	0.037	0.033	0.030	0.028	0.029	0.026
R11B	0.030	0.024	0.020	0.017	0.015	0.016	0.013
R12A	0.038	0.032	0.025	0.025	0.023	0.025	0.022
S11A	0.217	0.165	0.137	0.118	0.104	0.107	0.091
S11B	0.988	0.198	0.096	0.067	0.054	0.057	0.045
S12A	0.208	0.153	0.123	0.104	0.090	0.093	0.077
T11A	0.209	0.204	0.200	0.196	0.193	0.193	0.189
T11B	0.058	0.049	0.043	0.038	0.034	0.034	0.030
T12B	0.084	0.075	0.068	0.062	0.056	0.056	0.050
U11A	0.038	0.033	0.028	0.024	0.021	0.022	0.018
U11B	0.029	0.024	0.021	0.018	0.015	0.016	0.013
U12A	0.030	0.025	0.021	0.018	0.016	0.017	0.013
U12B	0.025	0.021	0.019	0.016	0.014	0.016	0.013
V11A	0.040	0.027	0.021	0.017	0.015	0.016	0.013
V11B	0.043	0.024	0.018	0.015	0.014	0.015	0.012
V12A	0.045	0.026	0.020	0.017	0.015	0.016	0.013
V12B	0.030	0.022	0.020	0.018	0.016	0.018	0.015
W11A	0.102	0.089	0.075	0.064	0.053	0.053	0.040
W11B	0.120	0.104	0.090	0.078	0.066	0.067	0.053
W12A	0.080	0.068	0.058	0.049	0.040	0.040	0.031
X11A	0.763	0.494	0.265	0.149	0.086	0.094	0.064
X11B	0.763	0.492	0.263	0.147	0.082	0.090	0.060

FNIR = FALSE NEGATIVE IDENT. RATE	N = NEUROTECHNOLOGY	P = SMU	Q = IRITECH	R = COGENT	S = SMARTSENSORS	T = CAMBRIDGE
FPIR = FALSE POSITIVE IDENT. RATE	U = L1	V = MORPHO	W = IRISID	X = CROSSMATCH	Y = KYNEN	

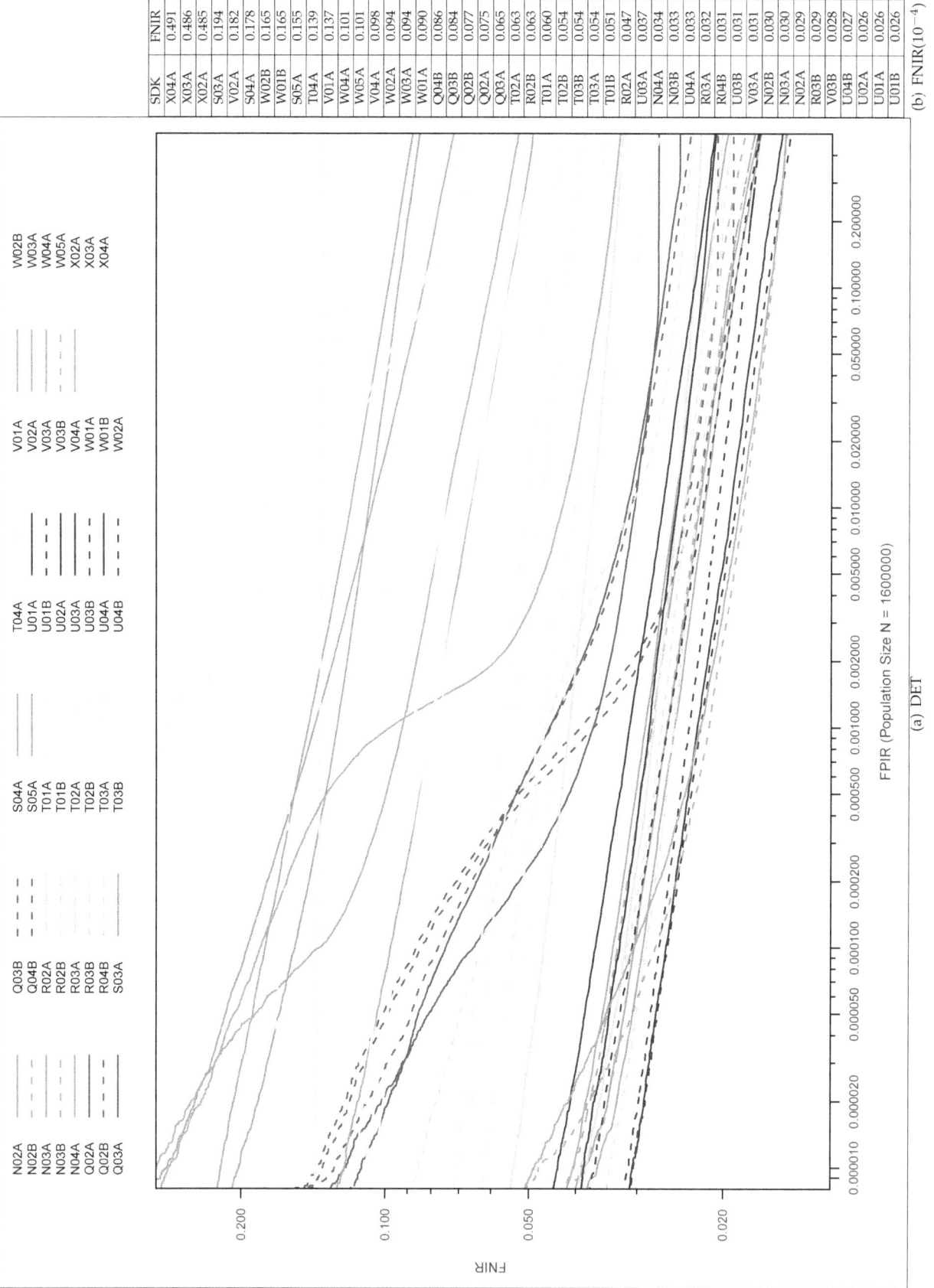

SDK	FNIR
X04A	0.491
X03A	0.486
X02A	0.485
S03A	0.194
V02A	0.182
S04A	0.178
W02B	0.165
W01B	0.165
S05A	0.155
T04A	0.139
V01A	0.137
W04A	0.101
W05A	0.101
V04A	0.098
W02A	0.094
W03A	0.094
W01A	0.090
Q04B	0.086
Q03B	0.084
Q02B	0.077
Q02A	0.075
Q03A	0.065
T02A	0.063
R02B	0.063
T01A	0.060
T02B	0.054
T03B	0.054
T03A	0.054
T01B	0.051
R02A	0.047
U03A	0.037
N04A	0.034
N03B	0.033
U04A	0.033
R03A	0.032
R04B	0.031
U03B	0.031
V03A	0.031
N02B	0.030
N03A	0.030
N02A	0.029
R03B	0.029
V03B	0.028
U04B	0.027
U02A	0.026
U01A	0.026
U01B	0.026

(b) FNIR(10^{-4})

(a) DET

FPIR (Population Size N = 1600000)

FNIR

Figure 18: For group 0 SDKs received from February to June 2011, detection-error tradeoff characteristic plotting FNIR vs FPIR for single-eye enrollment and search This is a 1:N DET - FPIR is logically N times an implied 1:1 FMR such that the left side of the DET corresponds to false match rates below 10^{-11} The size of enrolled population is $N = 1,600,000$ The two error rates are estimated over the large set S_{1b} The table elements are sorted and shaded light green when FNIR ≤ 0.05, and dark green for FNIR ≤ 0.025

N02A	Q03B	S04A	T04A	V01A	W02B
N02B	Q04B	S05A	U01A	V02A	W03A
N03A	R02A	T01A	U01B	V03A	W04A
N03B	R02B	T01B	U02A	V03B	W05A
N04A	R03A	T02A	U03A	V04A	X02A
Q02A	R03B	T02B	U03B	W01A	X03A
Q02B	R04B	T03A	U04A	W01B	X04A
Q03A	S03A	T03B	U04B	W02A	

FNIR = FALSE NEGATIVE IDENT. RATE	N = NEUROTECHNOLOGY
FPIR = FALSE POSITIVE IDENT. RATE	U = L1

P = SMU	Q = IRITECH	R = COGENT	S = SMARTSENSORS	T = CAMBRIDGE
V = MORPHO	W = IRISID	X = CROSSMATCH	Y = KYNEN	

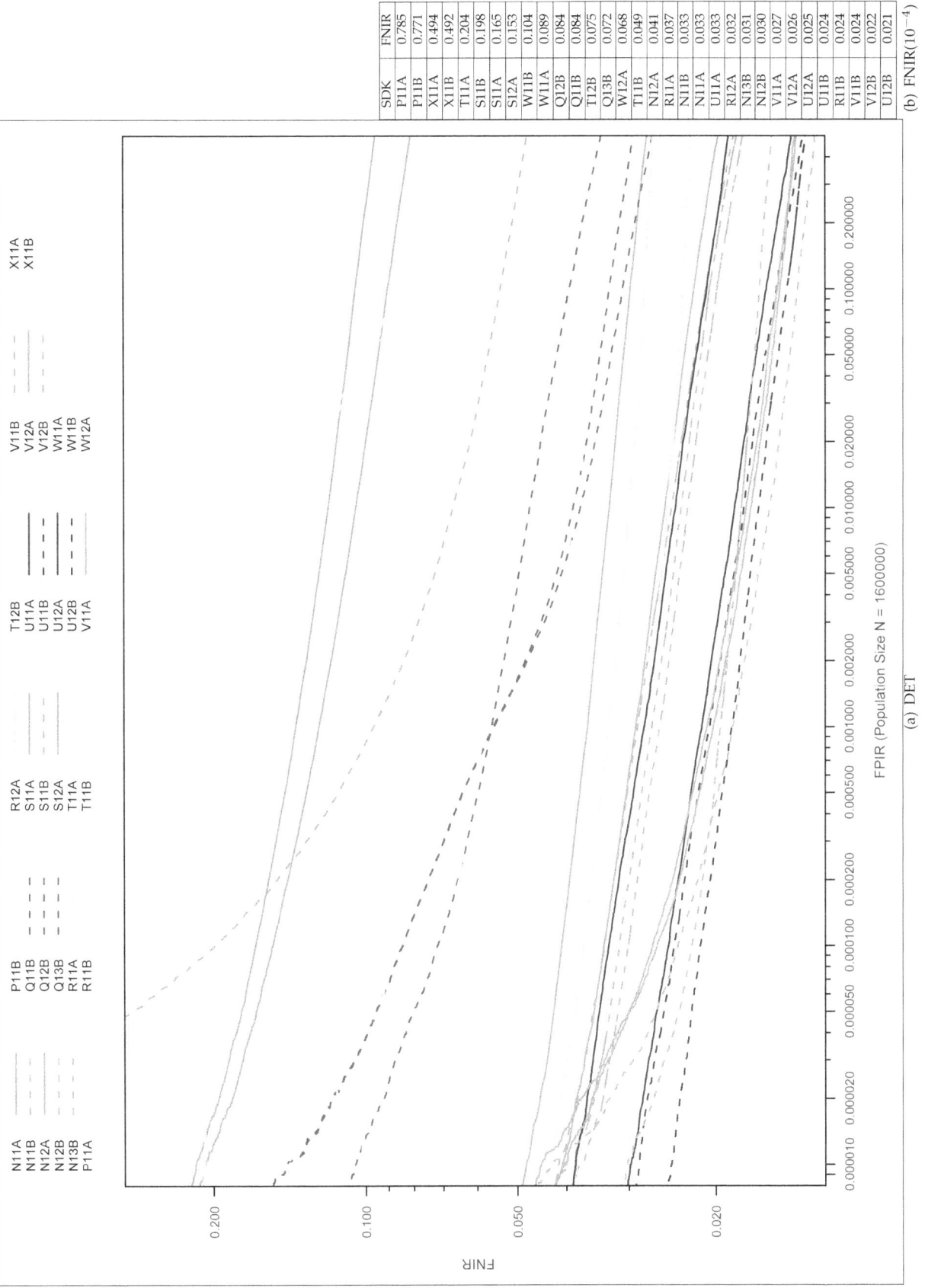

SDK	FNIR
P11A	0.785
P11B	0.771
X11A	0.494
X11B	0.492
T11A	0.204
S11B	0.198
S11A	0.165
S12A	0.153
W11B	0.104
W11A	0.089
Q12B	0.084
Q11B	0.084
T12B	0.075
Q13B	0.072
W12A	0.068
T11B	0.049
N12A	0.041
R11A	0.037
N11B	0.033
N11A	0.033
U11A	0.033
R12A	0.032
N13B	0.031
N12B	0.030
V11A	0.027
V12A	0.026
U12A	0.025
U11B	0.024
R11B	0.024
V11B	0.024
V12B	0.022
U12B	0.021

(b) FNIR(10^{-4})

(a) DET

FPIR (Population Size N = 1600000)

FNIR

Legend:

N11A	R12A	T12B
N11B	S11A	U11A
N12A	S11B	U11B
N12B	S12A	U12A
N13B	T11A	U12B
P11A	T11B	V11A

P11B	V11B	X11A
Q11B	V12A	X11B
Q12B	V12B	
Q13B	W11A	
R11A	W11B	
R11B	W12A	

Figure 19: For group 1 SDKs received in August 2011, detection-error tradeoff characteristic plotting FNIR vs FPIR for single-eye enrollment and search. This is a 1:N DET. DET - FPIR is logically N times an implied 1:1 FMR such that the left side of the DET corresponds to false match rates below 10^{-11} The size of enrolled population is $N = 1,600,000$ The two error rates are estimated over the large set S_{1b} The table elements are sorted and shaded light green when FNIR $\leq$ 0.05, and dark green for FNIR $\leq$ 0.025

FNIR = FALSE NEGATIVE IDENT. RATE	N = NEUROTECHNOLOGY	P = SMU	Q = IRITECH	R = COGENT	S = SMARTSENSORS	T = CAMBRIDGE
FPIR = FALSE POSITIVE IDENT. RATE	U = L1	V = MORPHO	W = IRISID	X = CROSSMATCH	Y = KYNEN	

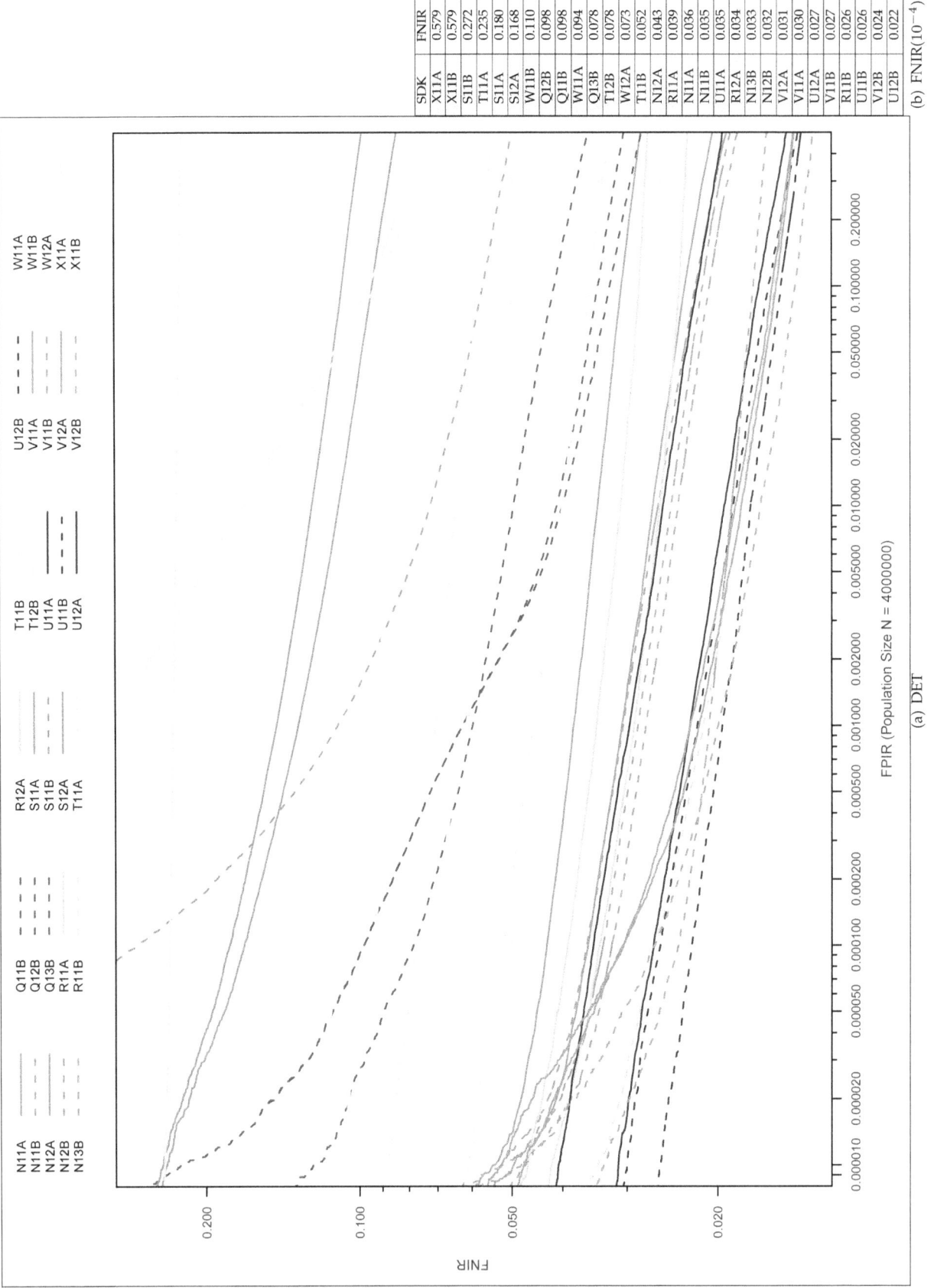

SDK	FNIR
X11A	0.579
X11B	0.579
S11B	0.272
T11A	0.235
S11A	0.180
S12A	0.168
W11B	0.110
Q12B	0.098
Q11B	0.098
W11A	0.094
Q13B	0.078
T12B	0.078
W12A	0.073
T11B	0.052
N12A	0.043
R11A	0.039
N11A	0.036
N11B	0.035
U11A	0.035
R12A	0.034
N13B	0.033
N12B	0.032
V12A	0.031
V11A	0.030
V12A	0.027
V11B	0.027
R11B	0.026
U11B	0.026
V12B	0.024
U12B	0.022

(b) FNIR(10^{-4})

(a) DET

Figure 20: For group 1 SDKs received in August 2011, detection-error tradeoff characteristic plotting FNIR vs FPIR for single-eye enrollment and search This is a 1:N DET - FPIR is logically N times an implied 1:1 FMR such that the left side of the DET corresponds to false match rates below 10^{-11} The size of enrolled population is $N = 3.9$ million The two error rates are estimated over the large set S_{16} The table elements are sorted and shaded light green when FNIR ≤ 0.05, and dark green for FNIR ≤ 0.025

FNIR = FALSE NEGATIVE IDENT. RATE	N = NEUROTECHNOLOGY	P = SMU	Q = IRITECH	R = COGENT	S = SMARTSENSORS	T = CAMBRIDGE
FPIR = FALSE POSITIVE IDENT. RATE	U = L1	V = MORPHO	W = IRISID	X = CROSSMATCH	Y = KYNEN	

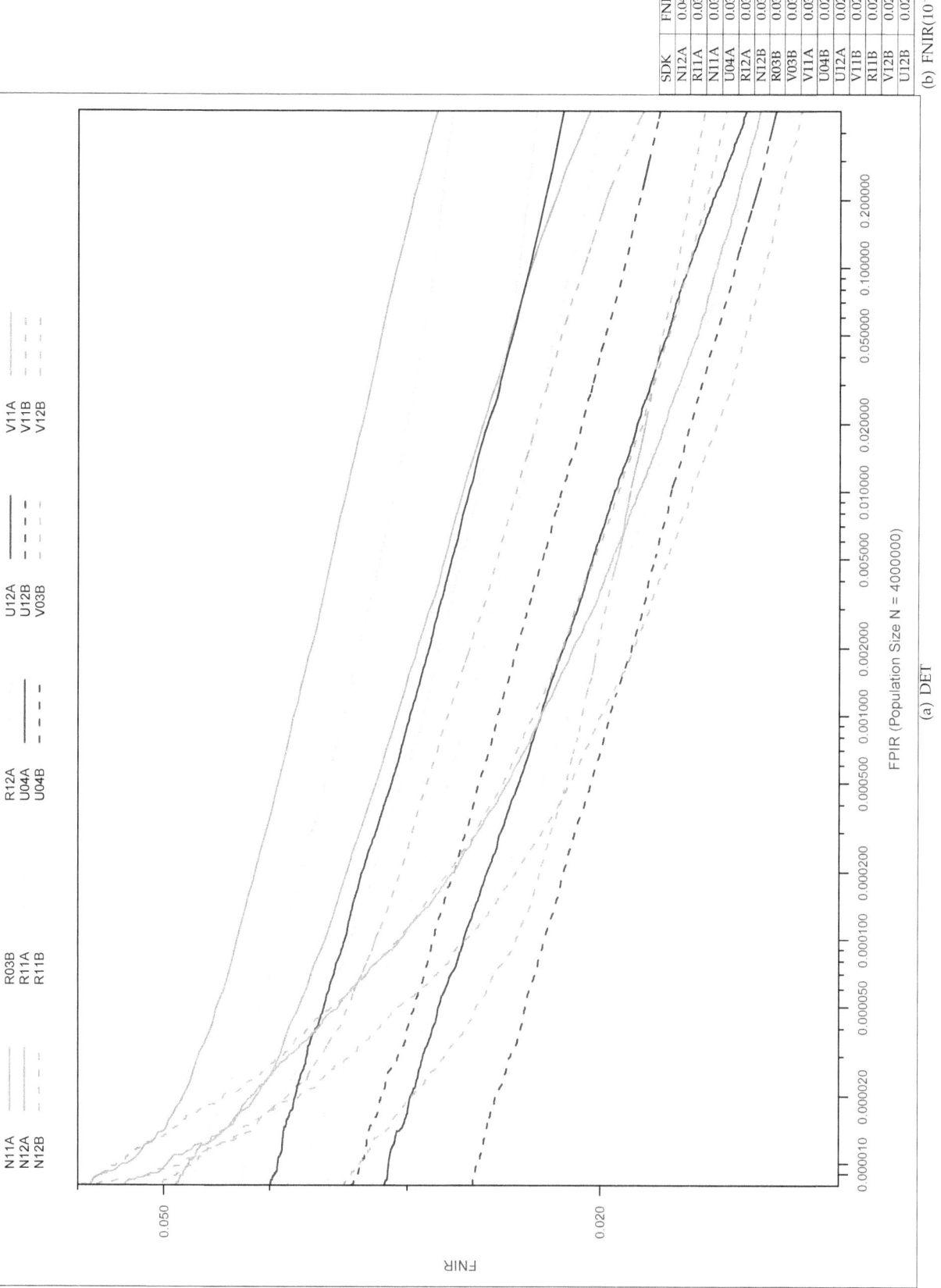

SDK	FNIR
N12A	0.043
R11A	0.039
N11A	0.036
U04A	0.034
R12A	0.034
N12B	0.032
R03B	0.031
V03B	0.030
V11A	0.030
U04B	0.028
U12A	0.027
V11B	0.027
R11B	0.026
V12B	0.024
U12B	0.022

(b) FNIR(10^{-4})

(a) DET

Figure 21: For the high accuracy SDKs, detection-error tradeoff characteristic plotting FNIR vs FPIR for single-eye enrollment and search. This is a 1:N DET - FPIR is logically N times an implied 1:1 FMR such that the left side of the DET corresponds to false match rates below 10^{-11}. The size of enrolled population is $N = 3.9$ million. The two error rates are estimated over the large set S_{1b}. The table elements are sorted and shaded light green when FNIR ≤ 0.05, and dark green for FNIR ≤ 0.025.

FNIR = FALSE NEGATIVE IDENT. RATE	N = NEUROTECHNOLOGY	P = SMU	Q = IRITECH	R = COGENT	S = SMARTSENSORS	T = CAMBRIDGE
FPIR = FALSE POSITIVE IDENT. RATE	U = L1	V = MORPHO	W = IRISID	X = CROSSMATCH	Y = KYNEN	

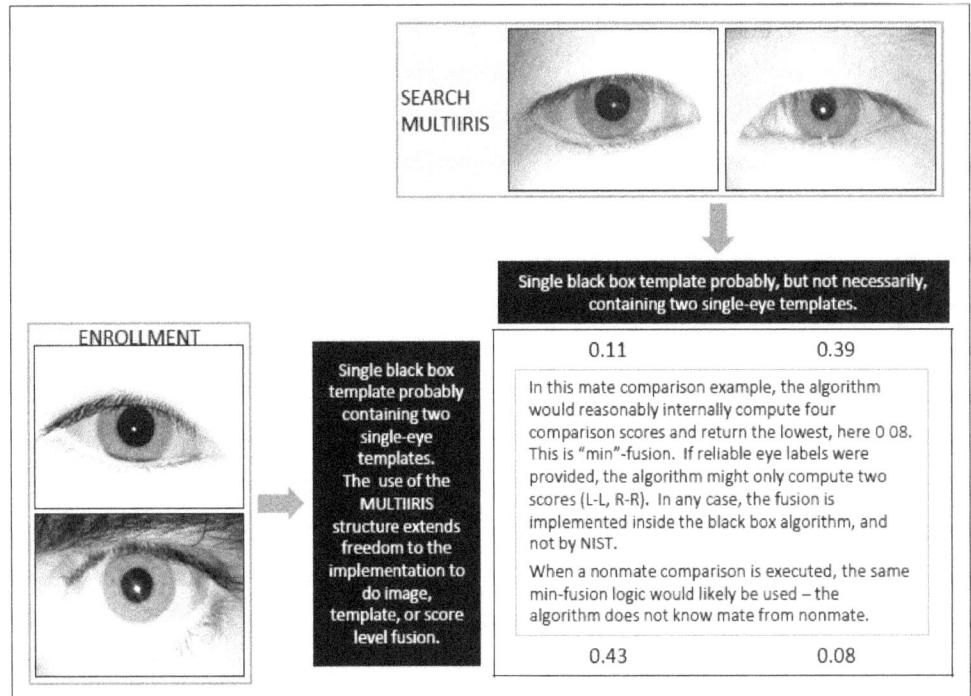

Figure 22: The encapsulation of images in a MULTIIRIS structure allows the implementation to exploit and fuse iris data as it sees fit. The simplest case is that the black box template (sizes are shown in Tables 6 and 7) embeds two single-eye templates and these are matched in the normal manner. However, more complicated schemes are possible. Regardless of the fusion scheme, the API requires exactly one scalar dissimilarity score per candidate.

6.6 Two-eye accuracy

Multiple biometric captures are known to afford improved recognition accuracy[29]. Iris texture has been reported to be largely epigenetic, such that extracted features from an individuals left and right irides are as uncorrelated as those from unrelated individuals [7]. This observation implies that the discriminative power of two irides is higher than one. However, as discussed below, when imaging is imperfect, both eyes may be affected. This section reports two-eye recognition performance.

6.6.1 Methods

The computation of accuracy and speed are identical to the single-eye case because the IREX III API document encapsulated both single-eye and two-eye cases in a single data structure that is passed to the template generator en-block. This template abstraction is depicted in Figure 22. Identification proceeds by generating a search template from one or more images, searching it against the enrollment database, and producing a list of candidates. Fusion is handled internally by the algorithm.

6.6.2 Results

The tabulations in Table 13 and plots of DETs in Figures 23 and 24 show that false negative identification error rates are lower for two-eye recognition than for single-eye recognition (Table 12, Figures 18 and 19). Broadly, the downward and leftward translations of the two-eye DETs show that accuracy is improved appreciably. Subsequent graphs and discussion (see section 6.8) clearly establish that at a fixed threshold, the translation is downward, corresponding to an improvement in FNIR. Particularly, the plot of the $FNIR_2$ vs. $FNIR_1$ in Figure 25 shows about a factor of two reduction in FNIR with some

(a) FNIR, Group 0

SDK	0.00001	0.0001	0.001	0.01	0.1
N02A	-	-	-	-	-
N02B	0.023	0.015	0.013	0.011	0.009
N03A	0.025	0.015	0.013	0.011	0.009
N03B	0.025	0.015	0.012	0.011	0.009
N04A	0.031	0.018	0.015	0.012	0.011
P02A	-	-	-	-	-
Q02A	0.086	0.056	0.034	0.020	0.016
Q02B	0.295	0.062	0.034	0.019	0.014
Q03A	0.120	0.056	0.026	0.019	0.016
Q03B	0.157	0.072	0.035	0.017	0.014
Q04B	0.161	0.073	0.037	0.017	0.013
R02A	0.031	0.028	0.025	0.023	0.021
R02B	-	-	-	-	-
R03A	0.035	0.033	0.031	0.029	0.028
R03B	0.020	0.015	0.013	0.011	0.010
R04B	0.023	0.017	0.013	0.011	0.010
S01B	-	-	-	-	-
S02B	-	-	-	-	-
S03A	0.229	0.129	0.087	0.061	0.044
S04A	0.156	0.119	0.096	0.078	0.065
S05A	0.148	0.108	0.085	0.071	0.062
T01A	0.036	0.029	0.025	0.021	0.018
T01B	0.034	0.027	0.023	0.019	0.015
T02A	0.037	0.030	0.026	0.022	0.019
T02B	0.034	0.028	0.024	0.020	0.017
T03A	0.034	0.028	0.024	0.020	0.016
T03B	0.035	0.029	0.024	0.020	0.016
T04A	0.078	0.075	0.073	0.071	0.070
U01A	0.019	0.015	0.013	0.011	0.009
U01B	0.019	0.016	0.013	0.011	0.009
U02A	0.019	0.015	0.013	0.011	0.009
U03A	0.027	0.022	0.019	0.016	0.013
U03B	0.023	0.019	0.016	0.013	0.011
U04A	0.019	0.017	0.015	0.013	0.012
U04B	0.015	0.013	0.012	0.010	0.009
V01A	0.264	0.130	0.055	0.042	0.032
V02A	0.215	0.131	0.074	0.022	0.016
V03A	0.039	0.022	0.014	0.010	0.008
V03B	0.039	0.020	0.013	0.010	0.008
V04A	0.076	0.053	0.041	0.032	0.025
W01A	0.979	0.979	0.979	0.979	0.875
W01B	0.085	0.075	0.068	0.062	0.057
W02A	0.054	0.045	0.037	0.030	0.024
W02B	0.085	0.075	0.068	0.062	0.057
W03A	0.054	0.044	0.036	0.030	0.024
W04A	0.055	0.047	0.039	0.032	0.026
W05A	0.055	0.047	0.039	0.032	0.026
X02A	-	-	-	-	-
X03A	-	-	-	-	-
X04A	-	-	-	-	-
Y02A	-	-	-	-	-
Y02B	-	-	-	-	-
Y03A	-	-	-	-	-
Y03B	-	-	-	-	-

(b) Rank 1, Group 0

SDK	$\beta(1)$	$\beta(20)$
N02A	0.009	0.007
N02B	0.010	0.007
N03A	0.010	0.007
N03B	0.010	0.007
N04A	0.012	0.008
P02A	0.419	0.401
Q02A	0.016	0.014
Q02B	0.016	0.012
Q03A	0.017	0.016
Q03B	0.016	0.014
Q04B	0.016	0.013
R02A	0.022	0.020
R02B	0.019	0.013
R03A	0.029	0.028
R03B	0.011	0.009
R04B	0.011	0.009
S01B	0.077	0.065
S02B	0.076	0.065
S03A	0.049	0.033
S04A	0.069	0.055
S05A	0.064	0.054
T01A	0.019	0.014
T01B	0.016	0.011
T02A	0.020	0.016
T02B	0.018	0.013
T03A	0.017	0.013
T03B	0.017	0.013
T04A	0.072	0.070
U01A	0.010	0.007
U01B	0.010	0.009
U02A	0.010	0.009
U03A	0.015	0.011
U03B	0.012	0.009
U04A	0.014	0.010
U04B	0.011	0.008
V01A	0.037	0.025
V02A	0.019	0.013
V03A	0.010	0.007
V03B	0.010	0.007
V04A	0.028	0.020
W01A	0.525	0.521
W01B	0.058	0.053
W02A	0.024	0.017
W02B	0.058	0.053
W03A	0.024	0.017
W04A	0.027	0.020
W05A	0.026	0.020
X02A	-	-
X03A	-	-
X04A	-	-
Y02A	-	-
Y02B	-	-
Y03A	-	-
Y03B	-	-

(c) FNIR, Group 1

	0.00001	0.0001	0.001	0.01	0.1
N11A	0.020	0.015	0.013	0.011	0.010
N11B	0.023	0.016	0.014	0.012	0.010
N12A	0.021	0.017	0.014	0.012	0.011
N12B	0.023	0.015	0.012	0.011	0.009
N13B	0.024	0.015	0.013	0.011	0.010
P11A	-	-	-	-	-
P11B	-	-	-	-	-
Q11B	0.231	0.066	0.041	0.023	0.018
Q12B	-	-	-	-	-
Q13B	0.114	0.051	0.035	0.027	0.022
R11A	0.021	0.017	0.015	0.014	0.013
R11B	0.020	0.015	0.011	0.009	0.008
R12A	0.020	0.016	0.014	0.013	0.012
S11A	0.159	0.115	0.093	0.079	0.070
S11B	-	-	-	-	-
S12A	0.161	0.112	0.089	0.074	0.065
T11A	0.121	0.113	0.107	0.102	0.098
T11B	0.029	0.024	0.020	0.017	0.014
T12B	0.059	0.046	0.038	0.030	0.023
U11A	0.024	0.020	0.017	0.014	0.011
U11B	0.018	0.015	0.012	0.010	0.009
U12A	0.018	0.015	0.012	0.010	0.007
U12B	0.015	0.013	0.011	0.010	0.008
V11A	0.031	0.019	0.013	0.010	0.008
V11B	0.036	0.017	0.012	0.009	0.007
V12A	0.037	0.020	0.013	0.010	0.008
V12B	0.016	0.012	0.011	0.010	0.008
W11A	0.051	0.042	0.035	0.028	0.023
W11B	0.057	0.048	0.040	0.033	0.027
W12A	0.104	0.100	0.096	0.093	0.090
X11A	0.861	0.611	0.318	0.179	0.104
X11B	-	-	-	-	-

(d) Rank 1, Group 1

SDK	$\beta(1)$	$\beta(20)$
N11A	0.011	0.008
N11B	0.010	0.008
N12A	0.012	0.010
N12B	0.010	0.007
N13B	0.010	0.008
P11A	0.682	0.122
P11B	-	-
Q11B	0.019	0.015
Q12B	0.549	0.546
Q13B	0.023	0.017
R11A	0.014	0.013
R11B	0.009	0.007
R12A	0.013	0.012
S11A	0.073	0.062
S11B	0.035	0.025
S12A	0.068	0.058
T11A	0.098	0.093
T11B	0.015	0.012
T12B	0.024	0.018
U11A	0.013	0.009
U11B	0.010	0.007
U12A	0.010	0.007
U12B	0.010	0.007
V11A	0.010	0.007
V11B	0.010	0.007
V12A	0.010	0.007
V12B	0.010	0.008
W11A	0.023	0.017
W11B	0.027	0.021
W12A	0.091	0.087
X11A	0.112	0.079
X11B	0.107	0.074

Table 13: FNIR at various FPIR values, for two-eye enrollment and search, estimated using set S_{2b}. The size of enrolled population is $N = 1,600,000$. The Table corresponds to Figures 23 and 24. The poor quality 330x330 images are not included. The narrow sub-tables give rank one miss rates (β values from equation 13). Cells are shaded light green when FNIR $\leq$ 0.025, and dark green for FNIR $\leq$ 0.01. A row containing hyphens indicates the large-set run did not complete successfully, typically due to an ability to handle large N, crashes, hangs, or timeouts. In these cases, β values are stated in blue for the small-set S_{2b} if available. Single eye results appear in Table 12

FNIR = FALSE NEGATIVE IDENT. RATE	N = NEUROTECHNOLOGY	P = SMU	Q = IRITECH	R = COGENT	S = SMARTSENSORS	T = CAMBRIDGE
FPIR = FALSE POSITIVE IDENT. RATE	U = L1	V = MORPHO	W = IRISID	X = CROSSMATCH	Y = KYNEN	

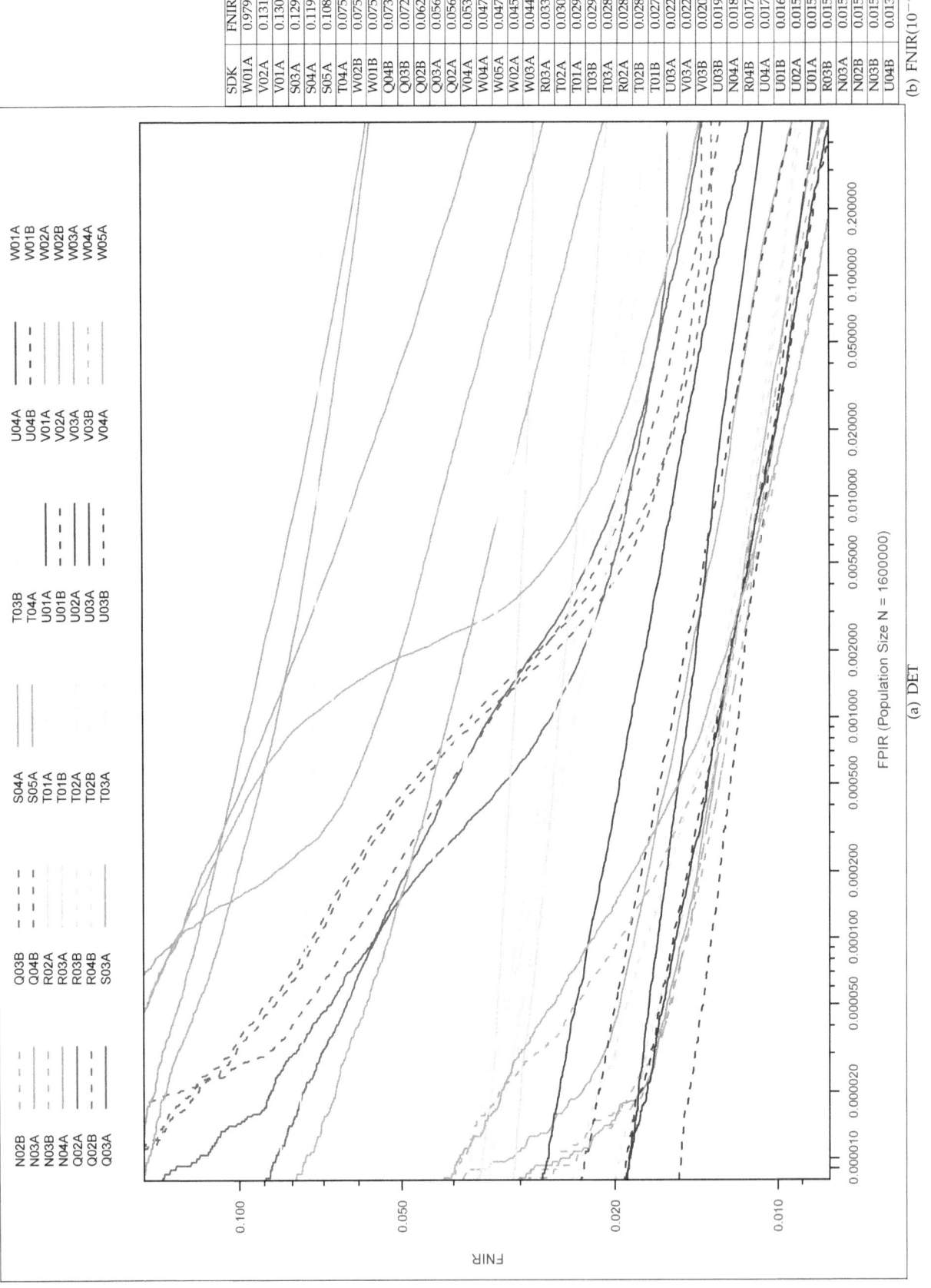

SDK	FNIR
W01A	0.979
V02A	0.131
V01A	0.130
S03A	0.129
S04A	0.119
S05A	0.108
T04A	0.075
W02B	0.075
W01B	0.075
Q04B	0.073
Q03B	0.072
Q02B	0.062
Q03A	0.056
Q02A	0.056
V04A	0.053
W04A	0.047
W05A	0.047
W02A	0.045
W03A	0.044
R03A	0.033
T02A	0.030
T01A	0.029
T03B	0.029
T03A	0.028
R02A	0.028
T02B	0.028
T01B	0.027
U03A	0.022
V03A	0.022
V03B	0.020
U03B	0.019
N04A	0.018
R04B	0.017
U04A	0.017
U01B	0.016
U02A	0.015
U01A	0.015
R03B	0.015
N03A	0.015
N02B	0.015
N03B	0.015
U04B	0.013

(b) FNIR(10^{-4})

(a) DET

Figure 23: For group 0 SDKs received from February to June 2011, two-eye detection-error tradeoff characteristic plotting FNIR vs FPIR The size of enrolled population is $N = 1,600,000$ The two error rates are estimated over the smaller search set S_{2b} The MULTIIRIS data structures contain two eyes for both enrollees and searches The table elements are sorted and shaded light green when FNIR ≤ 0.025

FNIR = FALSE NEGATIVE IDENT. RATE	N = NEUROTECHNOLOGY	P = SMU	Q = IRITECH	R = COGENT	S = SMARTSENSORS	T = CAMBRIDGE
FPIR = FALSE POSITIVE IDENT. RATE	U = L1	V = MORPHO	W = IRISID	X = CROSSMATCH	Y = KYNEN	

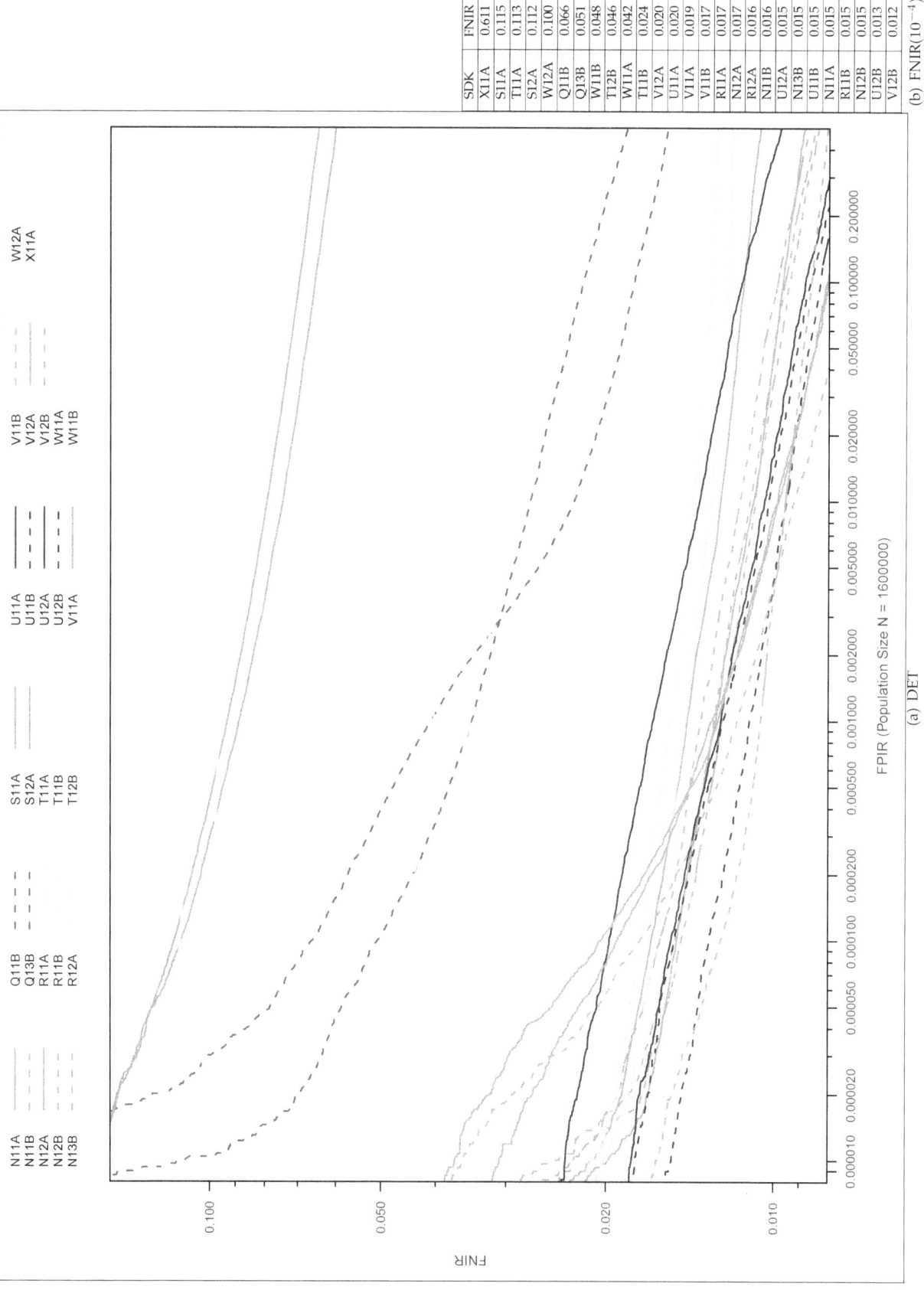

SDK	FNIR
X11A	0.611
S11A	0.115
T11A	0.113
S12A	0.112
W12A	0.100
Q11B	0.066
Q13B	0.051
W11B	0.048
T12B	0.046
W11A	0.042
T11B	0.024
V12A	0.020
U11A	0.020
V11A	0.019
V11B	0.017
R11A	0.017
N12A	0.017
R12A	0.016
N11B	0.016
U12A	0.015
N13B	0.015
U11B	0.015
N11A	0.015
R11B	0.015
N12B	0.015
U12B	0.013
V12B	0.012

(b) FNIR(10^{-4})

(a) DET

FPIR (Population Size N = 1600000)

FNIR

Figure 24: For group 1 SDKs received in August 2011, two-eye detection-error tradeoff characteristic plotting FNIR vs FPIR The size of enrolled population is $N = 1,600,000$ The two error rates are estimated over the smaller search set S_{2b} The MULTIIRIS data structures contain two eyes for both enrollees and searches The table elements are sorted and shaded light green when FNIR ≤ 0.025

| FNIR = FALSE NEGATIVE IDENT. RATE | N = NEUROTECHNOLOGY | P = SMU | Q = IRITECH | R = COGENT | S = SMARTSENSORS | T = CAMBRIDGE |
| FPIR = FALSE POSITIVE IDENT. RATE | U = L1 | V = MORPHO | W = IRISID | X = CROSSMATCH | Y = KYNEN | |

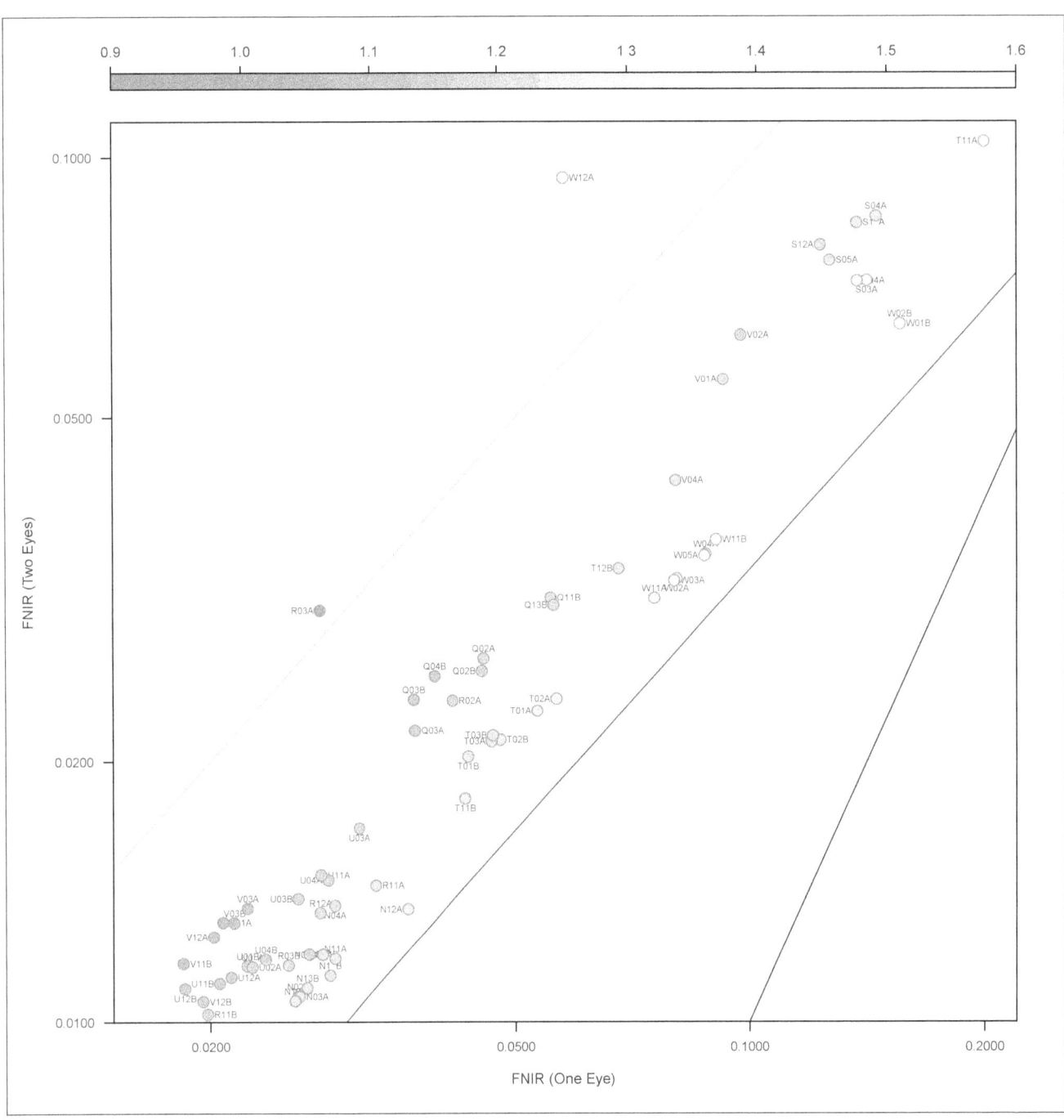

Figure 25: Two eye vs. single-eye FNIR at a threshold fixed to give FPIR= 0.001 in single-eye identification. The population size is N = 1,600,000. Each point corresponds to one algorithm. From left to right, the four straight lines plot $y = x$, $y = 2x/3$, $y = x/3$ and $y = x^2$. The first line reflects complete dependence of left and right eye FNIR and the last (best) line, complete independence. The points are color coded according to $\gamma = \log \text{FNIR}_1 / \log \text{FNIR}_2$.

FNIR = FALSE NEGATIVE IDENT. RATE	N = NEUROTECHNOLOGY	P = SMU	Q = IRITECH	R = COGENT	S = SMARTSENSORS	T = CAMBRIDGE
FPIR = FALSE POSITIVE IDENT. RATE	U = L1	V = MORPHO	W = IRISID	X = CROSSMATCH	Y = KYNEN	

variation between algorithms. Recalling that the implementation was presented with two unlabeled iris images and had to implement some fusion strategy, the plot reveals two things:

First, some providers appear to gain more from two eyes than others. This is evident from the clustering of providers' implementations along different paths in Figure 25. For example the W implementations exploit two-eye information more completely than, for example, Q, reaping greater reductions in FNIR.

However, these gains fall far short of that expected from independent samples. The fusion literature[1, 29] implies that left(L) and right(R) irides matched separately and fused with an effective scheme should yield

$$\text{FNMR}_2(\text{N}, \tau) = \text{FNMR}_L(\text{N}, \tau)\,\text{FNMR}_R(\text{N}, \tau) \tag{4}$$

where τ is the decision threshold and N is the population size. Under the assumption that left and right eyes match equally well, the equation simplifies to

$$\text{FNMR}_2(\text{N}, \tau) = \text{FNMR}_1(\text{N}, \tau)^2 \tag{5}$$

That this square dependency is not observed is indicative of a strong correlation in the quality of the samples involved in the fusion. Intuitively, we expect *bilateral* problems such as off-axis gaze, Ptosis (drooping eyelids)[30], Anridia (absence of iris), Coloboma (leading to mis-shaped pupils), Cataracts (opaque lens)[8, 28], oral medication or intoxicants that cause dilation, bright external ambient light causing reflections and pupil constriction, and blinking (affecting two-eye cameras) to degrade the quality of both samples.

Fusing both eyes typically increases the false positive rate by a factor of four over the single-eye case (see section 6.5), strongly suggesting an attempt by the algorithm to minimize false negatives (misses) by taking a "best-of" approach. That is, given four dissimilarity scores from two enrolled images matched against two search images, the lowest is taken. This is MIN fusion[23] and its implication, for false negative accuracy, is good because all four comparisons must fail (at some threshold) for a miss to occur. The implication for false positive rates is bad: any dissimilarity below the decision threshold will produce a false positive. The increase in false positive rates does not contradict the finding that extracted features from an individuals left and right irides are as uncorrelated as those from unrelated individuals [7]. Rather, the "best-of" configuration of the algorithms here reflects the importance of minimizing false negatives. Given different priories, the algorithms could be modified to require *both* left and right images to match; this making false positives more rare.

The correlation in FNIR accuracy is such that the FNIR typically drops by a factor of two when comparing two-eye accuracy to single eye accuracy at any fixed threshold. This does not appear to change appreciably depending on whether the camera captures one eye at a time, or both eyes concurrently.

Comparison with fingerprint: Figure 26 compares FNMR accuracy (at a fixed FMR) for one-finger vs. two-finger matching for algorithms submitted to the PFT evaluation. The fingerprints are left and right index captures from US-VISIT visa applicants. Although fingerprint matching was performed in 1:1 (verification) mode, and iris recognition was performed in 1:N (identification) mode, Figures 26 and 25 provide a direct comparison of the performance benefit to fusing irides vs. fusing fingerprints. The figures show that fingerprint fusion typically leads to a somewhat greater improvement in accuracy, specifically about a two-thirds reduction in FNIR for fingerprints compared to a factor of two reduction for iris. This is when likelihood ratio fusion is used for fingerprint matching. Figure 25 reveals that performing MAX-fusion for fingerprints (comparable to MIN-fusion for iris) is suboptimal.

6.7   Conclusion

As with fingerprint recognition based on left and right index fingers, the value of left and right iris images over a single image is not as much as that implied from independence assumptions. FNIR values drop linearly (by a factor of about two), rather than quadratically. This occurs because of the non-independent nature of left- and right-eye appearance, which itself arises from the *bilateral* nature of many eye acute and chronic diseases and non-ideal captures (even if the captures are sequential).

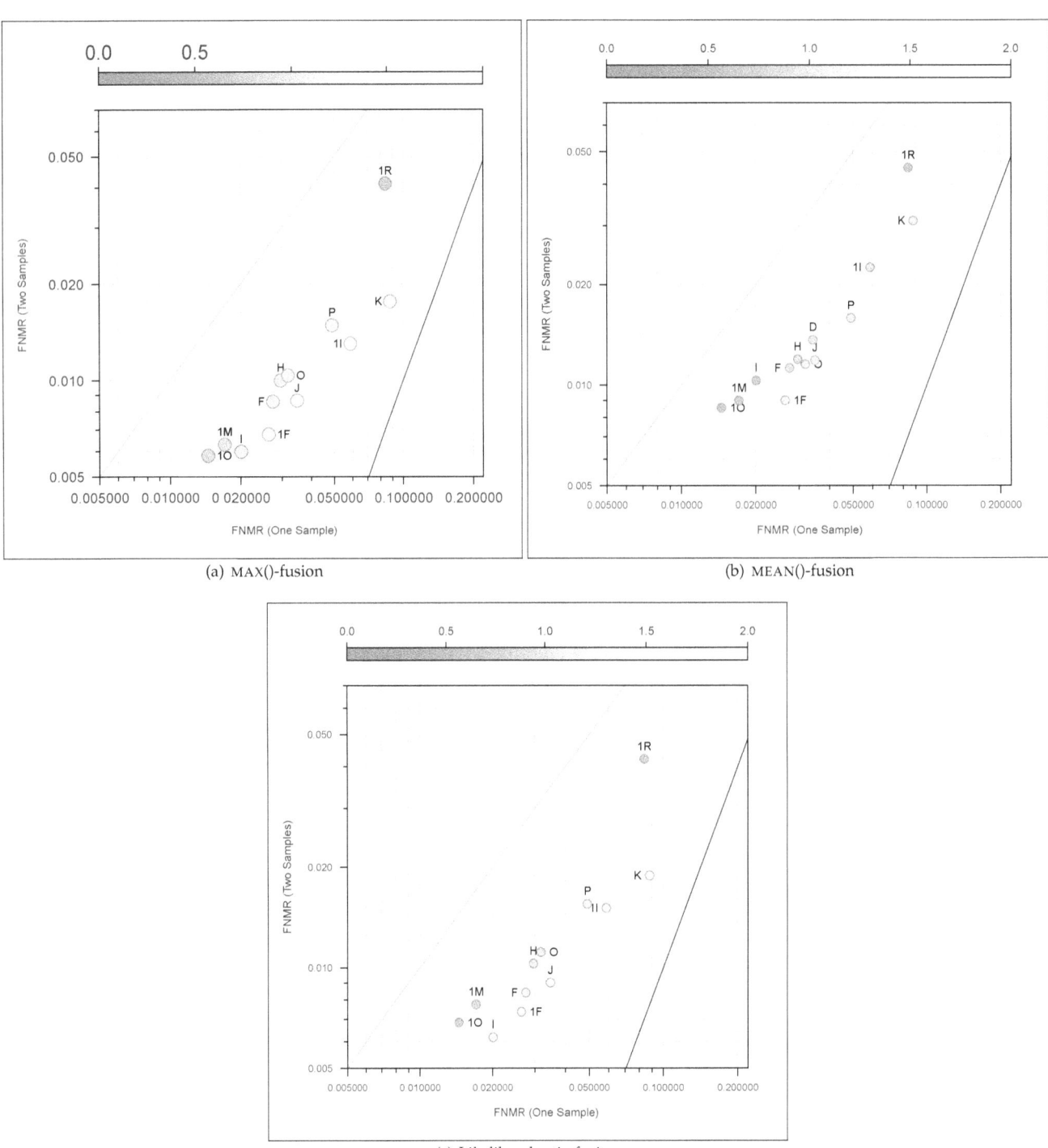

(a) MAX()-fusion

(b) MEAN()-fusion

(c) Likelihood ratio fusion

Figure 26: Two-finger vs. single-finger FNMR. For *mean* and *max* fusion rules the threshold is fixed to the single finger value that gives FMR= 0.001. For the likelihood ratio, the threshold is set to give FMR= 0.001 for single and two-fingers separately. Each point corresponds to a one-to-one fingerprint comparison algorithm submitted to NIST's PFT evaluation of fingerprint *verification* algorithms that use proprietary-template feature representations (not just standardized minutiae). The two lines, from left, are $y = x$ and $y = x^2$ indicating, respectively, dependence and independence of left and right finger failures.

FNIR = FALSE NEGATIVE IDENT. RATE	N = NEUROTECHNOLOGY	P = SMU	Q = IRITECH	R = COGENT	S = SMARTSENSORS	T = CAMBRIDGE
FPIR = FALSE POSITIVE IDENT. RATE	U = L1	V = MORPHO	W = IRISID	X = CROSSMATCH	Y = KYNEN	

6.8 Scalability

Scalability is defined here as the dependence of performance parameters on enrolled population size. This is important for applications which run for extended periods with the enrollment rate exceeding the un-enrollment rate. The specific measurements reported are the dependence of 1:N accuracy on N, and of the speed on N. Additionally the template sizes given in Tables 6 and 7 will drive the needed system memory and the number of computers required over time.

Note that there are limits to the applicability of IREX III results to applications in which population size is much larger than the 3.9M used here. These arise because the nonmate distribution is estimated empirically and projection to larger population sizes requires a *model* which inevitably requires validation.

6.8.1 Rationale

Biometric identification systems typically see net growth as individuals are added more frequently than they are deleted. Performance is known to degrade as enrolled population N increases because it is more likely to find biometrically similar ("lookalike") samples in a large population, and this calls for biometrics of high discriminatory power. The well known binomial model of biometric identification[1] holds that an N-person one-to-many identification implemented as N one-to-ones gives

$$\text{FPIR}(\text{N}, \tau) = 1 - (1 - \text{FMR}(\tau))^N \tag{6}$$

because each of the N comparisons has fixed probability of success equal to the false match rate $\text{FMR}(\tau)$, for threshold τ. For small FMR, this formula reduces (via the Taylor series) to the widely known linear form

$$\text{FPIR}(\text{N}, \tau) = \text{N} \ \text{FMR}(\tau) \tag{7}$$

The same theory indicates that false negative accuracy is independent of N because the genuine score is conventionally computed only as single one-to-one comparison independent of the other contents in the enrolled dataset, whence

$$\text{FNIR}(\text{N}, \tau) = \text{FNIR}(1, \tau) = \text{FNMR}(\tau). \tag{8}$$

The extension of this theory[13] to include rank effects is not needed here because rank is not part of this definition of FNIR. Rank-based performance is discussed in the next section.

An expensive core one-to-one comparison operation can be so expensive as to prohibit one-to-many search. For fingerprints and face[2], considerable engineering effort has been expended to expedite search using various partitioning, filtering, binning and multistage template approaches that apply the more accurate but slower algorithms only to a small portion of similar enrolled identities. This sometimes gives better than linear dependence on population size, N.

6.8.2 Results

The plots of Figure 27 show DET characteristics for two algorithms that typify the behaviors of identification algorithms under population growth. The notable observations are as follows. Analogous figures for all other algorithms exist in the IREX III APPENDICES.

▷ For most algorithms operating with a fixed threshold, the false positive rate grows linearly with population size, and the false negative identification rate is independent of population size. This is evident from the horizontal lines linking fixed threshold DET points in Figure 27(a). That figure shows the result for algorithm R03A. For most other algorithms equivalent behavior is presented in Appendix D of the IREX III APPENDICES.

▷ For some other algorithms (U04A, U04B, U12B, T11A, and V01A, V02A, V03A, V03B, V11B, V12B, V12B), the $\text{FNIR}(N, \tau)$ is not constant with population N and $\text{FPIR}(N, \tau)$ is, in some cases, independent of N. See the nearly vertical lines linking fixed threshold DET points in Figure 27(b). This is exactly contrary to the widely communicated model[1] that FPIR increases linearly with N (equation 7) and, FNIR is constant (equation 8). Why this occurs is a

FNIR = FALSE NEGATIVE IDENT. RATE	N = NEUROTECHNOLOGY	P = SMU	Q = IRITECH	R = COGENT	S = SMARTSENSORS	T = CAMBRIDGE
FPIR = FALSE POSITIVE IDENT. RATE	U = L1	V = MORPHO	W = IRISID	X = CROSSMATCH	Y = KYNEN	

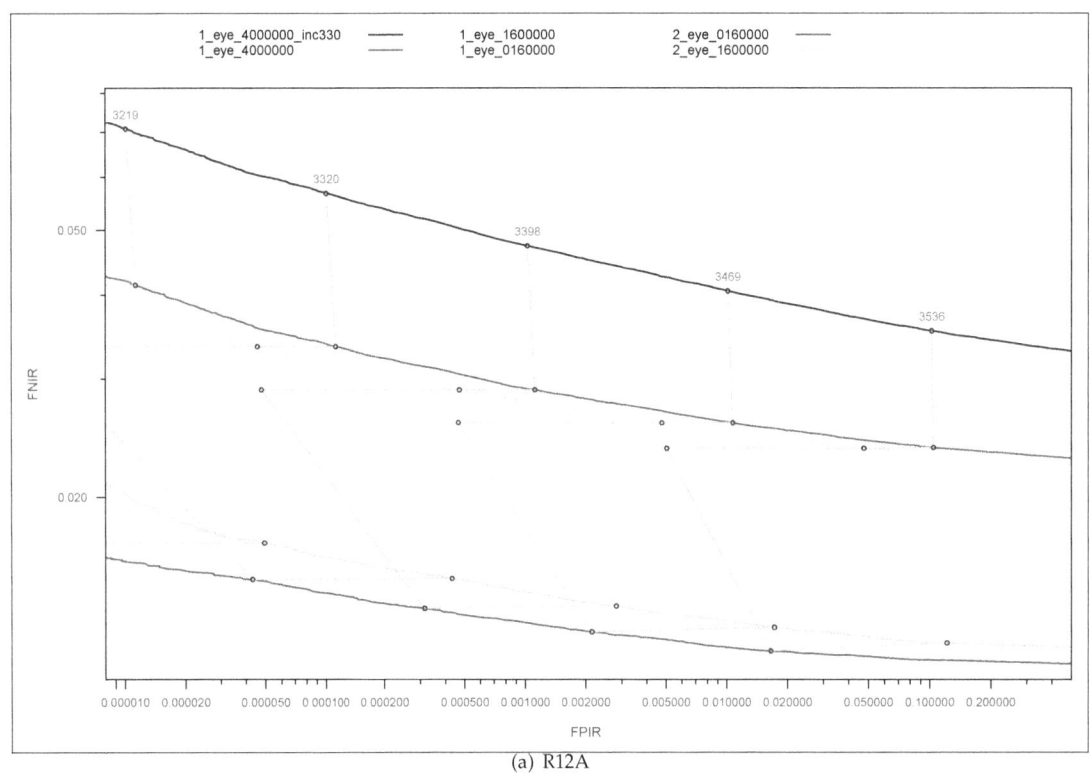

(a) R12A

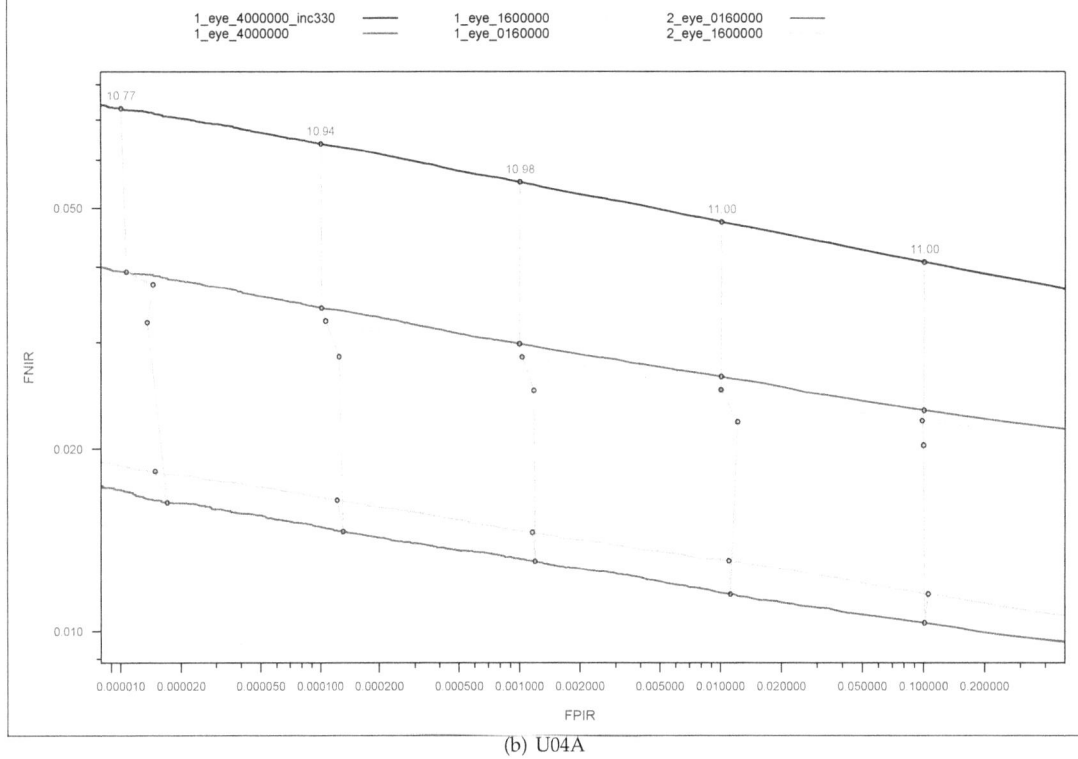

(b) U04A

Figure 27: **Scalability**: DETs for two algorithms executing searches from sets S_{nb} against enrolled populations of $N = (0.16, 1.6, 4)$ million (single-eye) and $N = (0.16, 1.6)$ million (two-eye). The straight lines link points of fixed threshold: Horizontal lines in the top figure indicate that enrolled population N is influential only on FPIR, via $\mathrm{FPIR}(N_2) = (N_2/N_1)\mathrm{FPIR}(N_1)$. Vertical lines in the lower figure show the algorithm is stabilizing FPIR under population size growth. Appendix C gives these figures for all algorithms: The behavior above is typical, and that shown below is present only for some U and V algorithms.

FNIR = FALSE NEGATIVE IDENT. RATE	N = NEUROTECHNOLOGY	P = SMU	Q = IRITECH	R = COGENT	S = SMARTSENSORS	T = CAMBRIDGE
FPIR = FALSE POSITIVE IDENT. RATE	U = L1	V = MORPHO	W = IRISID	X = CROSSMATCH	Y = KYNEN	

trade secret hidden inside the tested black box implementation. However, the effect is consistent with the use of gallery normlization[20] which has the effect of homogenizing the statistics of nonmate dissimilarity values across searches, and, evidently, across poulation sizes. The authors suggest this is beneficial in two ways:

- **Scalability**: Stabilization of FPIR under population growth takes the requirement to adjust thresholds out of the hands of system operators. This is important when N grows (when the rate of new enrollments exceeds that of deletions), and thresholds would ordinarily need to be increased.

- **Reduced FNIR**: For the algorithm pairs (U03A,U04A) and (U03B,U04B), the *03* algorithms have the traditional N-dependency (eq. 7) while the *04* algorithms exhibit the FPIR-stabilized behavior. However, the *04* algorithms also give fewer misses (see both FNIR and β in Table 12), without appreciable speed penalty (see Table 9).

6.9   Threshold calibration

System operators are tasked with adopting a threshold to implement security or policy objectives. In an identification system this will mean setting a threshold to achieve a low incidence of false positives.

Iris recognition has been attractive in this respect because there has been published discussion of how to set the threshold[21], in the form of false match rates being tabulated alongside thresholds[22], and these calibrations have theoretical support that has been published and peer reviewed[5]. Note that it is axiomatically true that arbitrarily low false match rates can be achieved with other biometric modalities by setting appropriately stringent thresholds. However, this practice is only tenable to the extent that false negative rates do not climb to unusable levels. This rests on the power of the biometric modality, the particular implementation, and the analog-to-digital imaging process.

6.9.1   Methods

This section gives empirical calibrations for the algorithms tested under IREX III, by tabulating the tails of the nonmate and mate distributions. By executing 311,427 nonmate searches against an enrolled population of 3,904,239 single eyes (see section 3.1), the algorithms are logically making 1216 billion nonmate comparisons[19]. This number compares with the 200 billion comparisons reported for the UAE study [21].

6.9.2   Results

Figure 28 shows threshold calibrations for two implementations. Analogous figures for all other algorithms exist in Appendix C of the IREX III APPENDICES. The notable observations are as follows.

▷ The T11B calibration implies fewer false matches than even the best SQRT-normalized calibration curve of the c. 2006 Cambridge algorithm applied to the 632,500 irides used in the UAE study[22]. For example at $\tau = 0.28$ with N = 3,904,239 the measured FPIR is 0.0001, implying via eq. 7, that FMR $= 2.56 \ 10^{-11}$. Figure 6 of the UAE study gives FMR $\approx 1 \ 10^{-9}$ which is a factor of about 40 higher than the current result. While the reason for this discrepancy is unknown, the following may be influential:

  - The Cambridge result[22] already includes a factor 7 elevation in FMR due to the multiple comparisons needed to handle in-plane rotation of the eyes. However, it is unreasonable to assume that similar rotation compensation methods were not present in the IREX III algorithms[20].

  - Algorithm improvements over the period 2006 to 2011,

---

[19]Technically the algorithms are tasked with reporting the 20 closest matches for each search, and may not, therefore, actually compute all comparisons. Such could occur if the search process included fast search strategies beyond exhaustive linear search.

[20]This is because the heavy use of single-eye cameras was known to the implementers; and single-eye cameras give higher variance in the in-plane rotation angle.

FNIR = FALSE NEGATIVE IDENT. RATE	N = NEUROTECHNOLOGY	P = SMU	Q = IRITECH	R = COGENT	S = SMARTSENSORS	T = CAMBRIDGE
FPIR = FALSE POSITIVE IDENT. RATE	U = LI	V = MORPHO	W = IRISID	X = CROSSMATCH	Y = KYNEN	

- The use of different imagery. Particularly many images used in IREX III have complete exposure of the iris because the eye is held open -see the example of Figure 2(a). This provides more information to the algorithm.

▷ For some algorithms, the calibration plots reveal an approximately linear dependency of log FPIR on threshold, τ, such that

$$\text{FPIR}(\tau) = ae^{b\tau} \tag{9}$$

Most algorithms also have a linear dependency on population size

$$\text{FPIR}(\tau) = N\,\text{FMR}(\tau) \tag{10}$$

for 1:1 false match rate, $\text{FMR}(\tau)$, and empirical constants a and b. In cases where the population size N is growing, FPIR can be maintained by reducing the threshold as follows. By equating equations 9 and 10 and differentiating with respect to time

$$abe^{b\tau}\frac{d\tau}{dt} = \text{FMR}\frac{dN}{dt} \tag{11}$$

whence

$$\frac{d\tau}{dt} = \frac{1}{bN}\frac{dN}{dt} \tag{12}$$

where it is assumed FMR is stable over time. This formula shows that, in this example, the threshold must increase at a rate proportional to the fractional increase in population size. This can be appreciable in rapidly expanding enrollments.

The threshold calibration plots are primarily useful in allowing thresholds to be set to purposefully target a false match rate. Importantly thresholds lower than those appearing in any of the graphs and tables here may be tenable if false negative performance is supported by other means. These are primarily good collection practices - see the IREX III FAILURE ANALYSIS SUPPLEMENTAL, and also automated and effective image quality asssesment implementations.

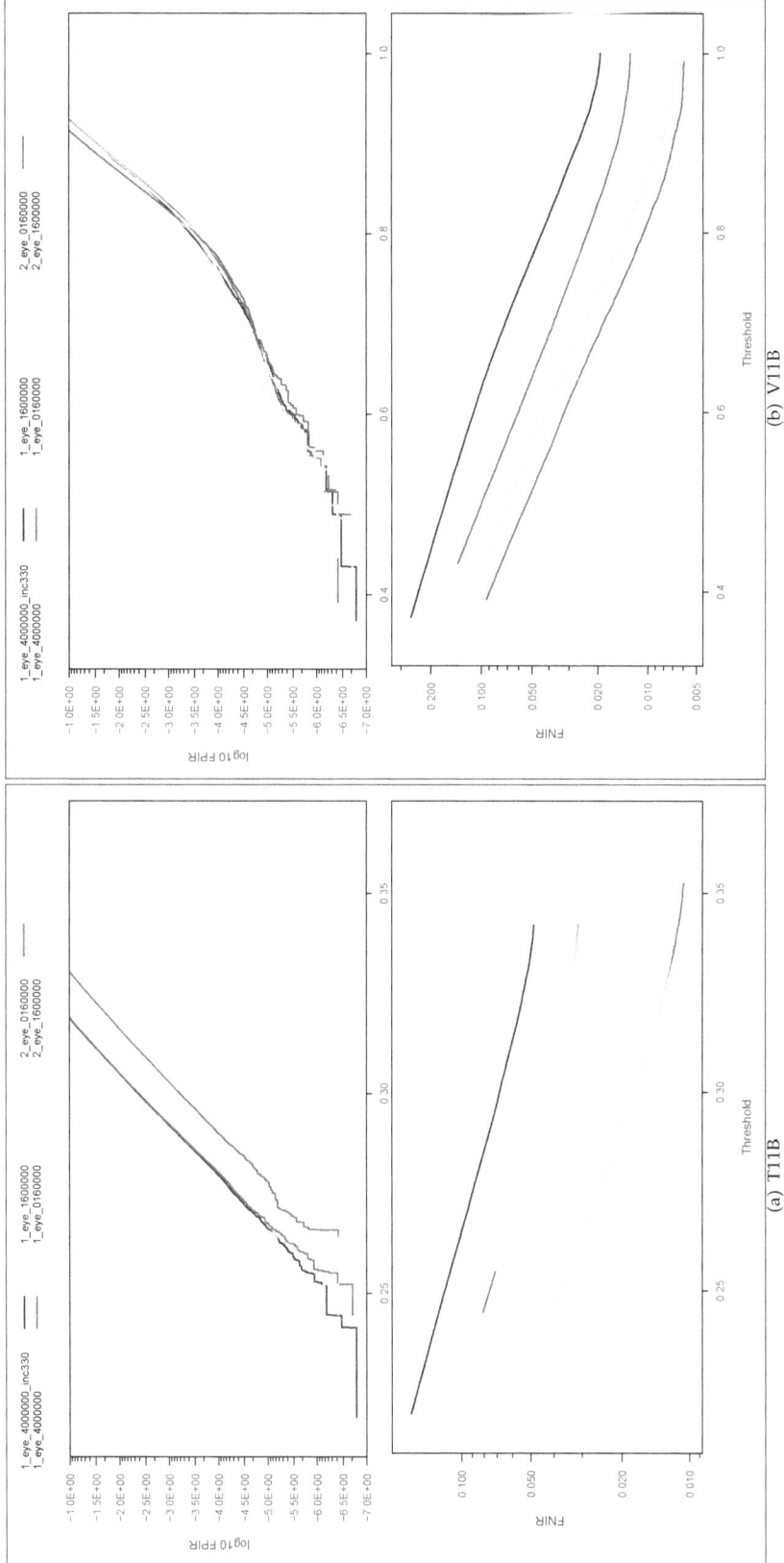

Figure 28: For two implementations, the plots show the left and right tails of, respectively, the nonmate (above) and mate distribution functions (below) The horizontal axis gives the dissimilarity threshold The vertical axes are logarithmic with, for FPIR, the labels being explicitly $\log_{10} \mathrm{FPIR}(\tau)$ For the Cambridge algorithm at left, the plot is related to Daugman's Figure 6 in [22] The various traces correspond to populations sizes $160000 \leq N \leq 3904239$ and one and two eyes These quantities were estimated over the large sets S_{1b} and S_{2b}

7   Rank-based accuracy

7.1   Rationale

This section presents metrics and results for evaluation of identification performance without any consideration of applying a threshold. This is particularly appropriate for *investigational* applications where the candidates returned on a potentially long candidate list would be inspected or adjudicated by an examiner. The distance or dissimilarity scores associated with each candidate would be used solely for sorting.

7.2   Metrics

For mated searches in an enrolled population of size, N, the metric appropriate for stating rank-based accuracy is termed the miss rate, $\beta(R, N)$ which is the fraction of searches for which the mate is not present on the L-entry candidate list at rank less than or equal to $1 \leq R \leq L$. This quantity is estimated over P searches for which there is an enrolled mate. Formally,

$$\beta(R, N) = 1 - \frac{1}{P} \sum_{p=1}^{P} \sum_{r=1}^{R} I_{pr} \tag{13}$$

where, for search in a population of size N, I_{pr} is 1 only if the identity of the r-th candidate is the same as the identity of search p, and 0 otherwise.

7.3   Results

Tables 14 and 15 show the rank 1 miss rates for all algorithms tested in the periods February-June 2011 and August-September 2011. The miss rates are shown for one and two-eye searching as a function of enrolled population sizes. In addition, they show the results with and without the pathologically compressed 330x330 images discussed in section 3.6. The graphs of Figure 29 and 30 give graphical depictions of the same.

The notable observations are as follows

    ▷ Rank 20 error rates are often, but not always, better than rank 1. This indicates that mates are sometimes present on the candidate lists at other than rank 1. This result does not reveal whether the rank 1 candidates in such cases are legitimate low-dissimilarity false matches or the result of ground-truth identity errors, or whether every candidate has high dissimilarity. To answer that question, the threshold based metric of section 6 has advantages because rank is not part of the accuracy definition.

    ▷ The cumulative match characteristics have positive slope, that is $\beta(20, N) > \beta(1, N)$. If the nonmate and mate distributions were separated this would not be true. In reality some mate scores are high enough (i.e. poor) that lower nonmate scores displace them from the top rank positions.

    ▷ Similarly $\beta(R, N)$ increases with N because lower dissimilarity nonmates are observed in the N-1 samples drawn from the parent nonmated distribution. The dependence is often linear in N, but a power-law model may be more appropriate. Note that nonmated comparisons harvested from mate searches should not be used to estimate FPIR; Instead, nonmated searches should be performed.

(a) All images, Group 0

	Eyes=1 0020000	Eyes=1 0160000	Eyes=1 1600000	Eyes=1 4000000	Eyes=2 0020000	Eyes=2 0160000	Eyes=2 1600000
N02A	0.037	0.044	0.050	-	0.018	0.021	0.027
N02B	0.037	0.044	0.051	0.054	0.018	0.021	0.026
N03A	0.038	0.044	0.051	0.054	0.018	0.022	0.027
N03B	0.036	0.044	0.051	0.054	0.017	0.020	0.025
N04A	0.040	0.048	0.056	0.060	0.022	0.027	0.035
P02A	0.146	0.164	0.177	-	0.437	0.447	0.459
Q02A	0.034	0.040	0.045	0.048	0.019	0.022	0.027
Q02B	0.030	0.038	0.044	0.047	0.018	0.021	0.027
Q03A	0.036	0.042	0.047	0.050	0.021	0.023	0.029
Q03B	0.026	0.033	0.039	0.043	0.017	0.020	0.027
Q04B	0.025	0.032	0.039	0.043	0.016	0.020	0.028
R02A	0.033	0.041	0.046	0.049	0.020	0.023	0.026
R02B	0.028	0.033	0.038	0.041	0.016	0.017	0.024
R03A	0.025	0.033	0.036	0.039	0.029	0.031	0.033
R03B	0.025	0.028	0.032	0.034	0.012	0.013	0.016
R04B	0.024	0.028	0.032	0.035	0.011	0.013	0.016
S01B	0.053	0.066	0.081	-	0.072	0.081	0.095
S02B	0.051	0.062	0.076	-	0.073	0.080	0.092
S03A	0.079	0.097	0.117	0.126	0.042	0.052	0.068
S04A	0.092	0.108	0.127	0.135	0.060	0.069	0.084
S05A	0.092	0.105	0.117	0.123	0.060	0.066	0.076
T01A	0.063	0.069	0.075	0.078	0.030	0.034	0.039
T01B	0.064	0.073	0.079	0.082	0.041	0.043	0.049
T02A	0.078	0.086	0.091	0.094	0.041	0.045	0.050
T02B	0.073	0.081	0.086	0.088	0.044	0.046	0.050
T03A	0.044	0.052	0.059	0.063	0.020	0.024	0.030
T03B	0.044	0.052	0.059	0.063	0.020	0.024	0.030
T04A	0.186	0.187	0.188	0.189	0.111	0.110	0.112
U01A	0.021	0.027	0.032	0.035	0.011	0.014	0.018
U01B	0.020	0.026	0.031	0.034	0.011	0.014	0.018
U02A	0.021	0.027	0.032	0.035	0.011	0.014	0.018
U03A	0.029	0.036	0.042	0.046	0.016	0.020	0.025
U03B	0.025	0.031	0.037	0.040	0.013	0.017	0.022
U04A	0.029	0.036	0.042	0.046	0.014	0.017	0.022
U04B	0.025	0.025	0.037	0.040	0.012	0.015	0.018
V01A	0.065	0.078	0.094	-	0.034	0.044	0.056
V02A	0.042	0.050	0.058	0.058	0.020	0.025	0.030
V03A	0.020	0.025	0.030	0.032	0.010	0.013	0.017
V03B	0.018	0.023	0.027	0.029	0.009	0.011	0.015
V04A	0.060	0.071	0.082	0.088	0.027	0.034	0.043
W01A	0.064	0.070	0.076	0.685	0.027	0.030	0.525
W01B	0.163	0.166	0.171	0.173	0.070	0.072	0.076
W02A	0.049	0.062	0.074	0.079	0.021	0.027	0.035
W02B	0.163	0.166	0.171	0.174	0.070	0.072	0.077
W03A	0.049	0.061	0.073	0.078	0.021	0.027	0.034
W04A	0.061	0.073	0.084	0.089	0.024	0.031	0.038
W05A	0.061	0.073	0.084	0.088	0.024	0.031	0.037
X02A	0.121	0.138	0.167	-	0.129	0.145	-
X03A	0.077	0.100	0.127	-	0.093	0.117	-
X04A	0.072	0.095	0.124	-	0.088	0.112	-
Y02A	-	-	-	-	-	-	-
Y02B	0.151	-	-	-	0.107	-	-
Y03A	0.134	0.149	-	-	0.066	0.078	-
Y03B	0.116	0.134	-	-	0.088	-	-

(b) Excluding 330x330, Group 0

	Eyes=1 0020000	Eyes=1 0160000	Eyes=1 1600000	Eyes=1 4000000	Eyes=2 0020000	Eyes=2 0160000	Eyes=2 1600000
N02A	0.014	0.016	0.018	-	0.007	0.008	0.009
N02B	0.014	0.016	0.019	0.020	0.007	0.008	0.010
N03A	0.013	0.016	0.018	0.020	0.007	0.008	0.010
N03B	0.015	0.017	0.020	0.022	0.007	0.008	0.010
N04A	0.016	0.019	0.022	0.024	0.008	0.009	0.012
P02A	0.135	0.148	0.161	-	0.399	0.408	0.419
Q02A	0.021	0.025	0.028	0.029	0.012	0.013	0.016
Q02B	0.020	0.024	0.027	0.029	0.012	0.013	0.016
Q03A	0.023	0.026	0.029	0.030	0.014	0.015	0.017
Q03B	0.016	0.019	0.023	0.025	0.011	0.012	0.016
Q04B	0.015	0.018	0.022	0.024	0.010	0.011	0.016
R02A	0.027	0.032	0.034	0.036	0.018	0.020	0.022
R02B	0.021	0.024	0.027	0.029	0.009	0.013	0.013
R03A	0.018	0.023	0.025	0.026	0.028	0.028	0.029
R03B	0.018	0.020	0.022	0.023	0.009	0.009	0.011
R04B	0.018	0.020	0.022	0.024	0.009	0.009	0.011
S01B	0.039	0.049	0.058	-	0.062	0.068	0.077
S02B	0.037	0.045	0.054	-	0.064	0.068	0.076
S03A	0.060	0.075	0.090	0.097	0.029	0.038	0.049
S04A	0.078	0.090	0.104	0.111	0.052	0.059	0.069
S05A	0.076	0.087	0.096	0.101	0.052	0.057	0.064
T01A	0.033	0.039	0.043	0.045	0.013	0.016	0.019
T01B	0.021	0.027	0.031	0.034	0.010	0.013	0.016
T02A	0.038	0.043	0.047	0.049	0.015	0.017	0.020
T02B	0.030	0.035	0.038	0.040	0.013	0.015	0.018
T03A	0.027	0.032	0.036	0.038	0.013	0.014	0.017
T03B	0.027	0.032	0.036	0.038	0.013	0.014	0.017
T04A	0.137	0.136	0.137	0.137	0.072	0.070	0.072
U01A	0.013	0.015	0.018	0.019	0.007	0.008	0.010
U01B	0.012	0.015	0.017	0.019	0.007	0.008	0.010
U02A	0.013	0.015	0.018	0.019	0.007	0.008	0.010
U03A	0.018	0.021	0.024	0.026	0.010	0.012	0.015
U03B	0.015	0.017	0.020	0.022	0.008	0.010	0.012
U04A	0.018	0.021	0.024	0.026	0.009	0.011	0.014
U04B	0.015	0.017	0.020	0.022	0.008	0.009	0.011
V01A	0.048	0.055	0.066	-	0.024	0.029	0.037
V02A	0.030	0.034	0.039	0.039	0.014	0.016	0.019
V03A	0.014	0.015	0.017	0.019	0.007	0.008	0.010
V03B	0.014	0.015	0.017	0.019	0.007	0.008	0.010
V04A	0.044	0.051	0.059	0.063	0.018	0.021	0.028
W01A	0.050	0.053	0.057	0.685	0.021	0.021	0.525
W01B	0.136	0.139	0.143	0.145	0.053	0.055	0.058
W02A	0.038	0.047	0.056	0.060	0.016	0.019	0.024
W02B	0.136	0.139	0.143	0.145	0.053	0.055	0.058
W03A	0.038	0.047	0.055	0.059	0.016	0.019	0.024
W04A	0.048	0.056	0.065	0.069	0.019	0.022	0.027
W05A	0.048	0.056	0.065	0.068	0.018	0.022	0.026
X02A	0.091	0.105	0.128	-	0.101	0.113	-
X03A	0.057	0.072	0.093	-	0.073	0.089	-
X04A	0.052	0.068	0.090	-	0.069	0.085	-
Y02A	-	-	-	-	-	-	-
Y02B	0.112	-	-	-	0.074	-	-
Y03A	0.116	0.128	-	-	0.055	0.063	-
Y03B	0.085	0.097	-	-	0.057	-	-

Table 14: For group 0 SDKs received from February to June 2011, rank one miss rates, $\beta(1)$, for recognition of one and two eyes as a function of enrolled population size. The estimates are computed over the mated searches from the large one- and two-eye sets, S_{1b} and S_{2b}, when that result is available and $N \geq 160{,}000$, or from the smaller sets S_{xa}, otherwise, in which case the text appears blue. The small set is only used when the large set run could not be completed. Hyphens indicate that no run could be completed. Cells are shaded light and dark green when the miss rates are less than or equal to 0.05 and 0.025 and 0.025 (single-eye), and 0.025 and 0.01 (two-eye), respectively.

FNIR = FALSE NEGATIVE IDENT. RATE	N = NEUROTECHNOLOGY	P = SMU	Q = IRITECH	R = COGENT	S = SMARTSENSORS	T = CAMBRIDGE
FPIR = FALSE POSITIVE IDENT. RATE	U = L1	V = MORPHO	W = IRISID	X = CROSSMATCH	Y = KYNEN	

(a) All images, Group 1

	Number of eyes = 1				Number of eyes = 2		
	0020000	0160000	1600000	4000000	0020000	0160000	1600000
N11A	0.040	0.048	0.056	0.059	0.019	0.023	0.029
N11B	0.036	0.044	0.051	0.055	0.018	0.021	0.026
N12A	0.049	0.056	0.063	0.067	0.021	0.025	0.031
N12B	0.055	0.062	0.068	0.070	0.032	0.035	0.040
N13B	0.055	0.063	0.068	0.071	0.031	0.035	0.040
P11A	0.146	0.166	0.217	-	0.129	0.143	0.708
P11B	-	0.218	0.266	-	-	-	-
Q11B	0.034	0.042	0.048	0.052	0.018	0.022	0.029
Q12B	0.031	0.039	0.046	0.050	0.016	0.021	0.550
Q13B	0.038	0.047	0.055	0.058	0.019	0.025	0.032
R11A	0.031	0.038	0.042	0.044	0.014	0.017	0.020
R11B	0.018	0.022	0.026	0.028	0.009	0.011	0.015
R12A	0.025	0.033	0.036	0.039	0.014	0.016	0.018
S11A	0.100	0.114	0.127	0.133	0.069	0.074	0.085
S11B	0.055	0.065	0.077	0.083	0.030	0.037	0.049
S12A	0.083	0.098	0.111	0.117	0.063	0.069	0.079
T11A	0.151	0.176	0.221	0.254	0.072	0.087	0.120
T11B	0.042	0.049	0.055	0.058	0.018	0.021	0.026
T12B	0.064	0.071	0.078	0.082	0.027	0.030	0.037
U11A	0.026	0.031	0.037	0.040	0.013	0.017	0.022
U11B	0.018	0.023	0.029	0.031	0.010	0.013	0.017
U12A	0.019	0.024	0.029	0.032	0.011	0.013	0.017
U12B	0.018	0.023	0.029	0.031	0.010	0.013	0.017
V11A	0.018	0.023	0.028	0.031	0.010	0.012	0.016
V11B	0.016	0.020	0.024	0.026	0.009	0.010	0.014
V12A	0.017	0.021	0.025	0.028	0.009	0.011	0.015
V12B	0.021	0.027	0.032	0.035	0.010	0.013	0.017
W11A	0.046	0.059	0.070	0.075	0.020	0.026	0.033
W11B	0.063	0.075	0.086	0.090	0.025	0.031	0.039
W12A	0.037	0.046	0.056	0.060	0.016	0.021	0.115
X11A	0.077	0.098	0.125	0.139	0.092	0.115	0.146
X11B	0.071	0.093	0.121	0.136	0.087	0.106	0.139

(b) Excluding 330x330, Group 1

	Number of eyes = 1				Number of eyes = 2		
	0020000	0160000	1600000	4000000	0020000	0160000	1600000
N11A	0.016	0.019	0.021	0.023	0.008	0.009	0.011
N11B	0.015	0.018	0.020	0.022	0.008	0.008	0.010
N12A	0.024	0.027	0.029	0.031	0.009	0.010	0.012
N12B	0.014	0.017	0.019	0.020	0.007	0.008	0.010
N13B	0.014	0.017	0.019	0.021	0.007	0.008	0.010
P11A	0.135	0.150	0.193	-	0.120	0.130	0.682
P11B	-	0.123	0.177	-	-	-	-
Q11B	0.025	0.030	0.034	0.036	0.014	0.016	0.019
Q12B	0.023	0.027	0.031	0.034	0.012	0.015	0.549
Q13B	0.029	0.034	0.039	0.042	0.015	0.019	0.023
R11A	0.023	0.027	0.029	0.031	0.011	0.013	0.014
R11B	0.012	0.014	0.016	0.017	0.006	0.007	0.009
R12A	0.018	0.023	0.025	0.026	0.011	0.012	0.013
S11A	0.085	0.097	0.107	0.112	0.061	0.065	0.073
S11B	0.042	0.049	0.057	0.061	0.024	0.028	0.035
S12A	0.070	0.082	0.093	0.098	0.057	0.061	0.068
T11A	0.128	0.149	0.193	0.224	0.057	0.069	0.098
T11B	0.027	0.031	0.034	0.036	0.011	0.012	0.015
T12B	0.046	0.052	0.056	0.059	0.018	0.020	0.024
U11A	0.016	0.019	0.022	0.024	0.008	0.010	0.013
U11B	0.011	0.014	0.016	0.017	0.007	0.008	0.010
U12A	0.012	0.014	0.017	0.018	0.007	0.008	0.010
U12B	0.011	0.014	0.016	0.017	0.007	0.008	0.010
V11A	0.012	0.014	0.016	0.018	0.007	0.008	0.010
V11B	0.012	0.013	0.015	0.016	0.007	0.007	0.010
V12A	0.012	0.014	0.016	0.018	0.007	0.008	0.010
V12B	0.013	0.015	0.018	0.019	0.007	0.008	0.010
W11A	0.036	0.044	0.053	0.056	0.016	0.019	0.023
W11B	0.049	0.058	0.067	0.070	0.020	0.023	0.027
W12A	0.028	0.034	0.040	0.043	0.013	0.015	0.091
X11A	0.058	0.073	0.094	0.104	0.073	0.089	0.112
X11B	0.053	0.069	0.090	0.101	0.069	0.083	0.107

Table 15: For group 1 SDKs received in August 2011, rank one miss rates, $\beta(1)$, for recognition of one and two eyes as a function of enrolled population size. The estimates are computed over the mated searches from the large one- and two-eye sets, S_{1b} and S_{2b}, when that result is available and $N \geq 160,000$, or from the smaller sets S_{xa}, otherwise, in which case the text appears blue. The small set is only used when the large set run could not be completed. Hyphens indicate that no run could be completed. Cells are shaded light and dark green when the miss rates are less than or equal to 0.05 and 0.025 (single-eye), and 0.025 and 0.01 (two-eye), respectively.

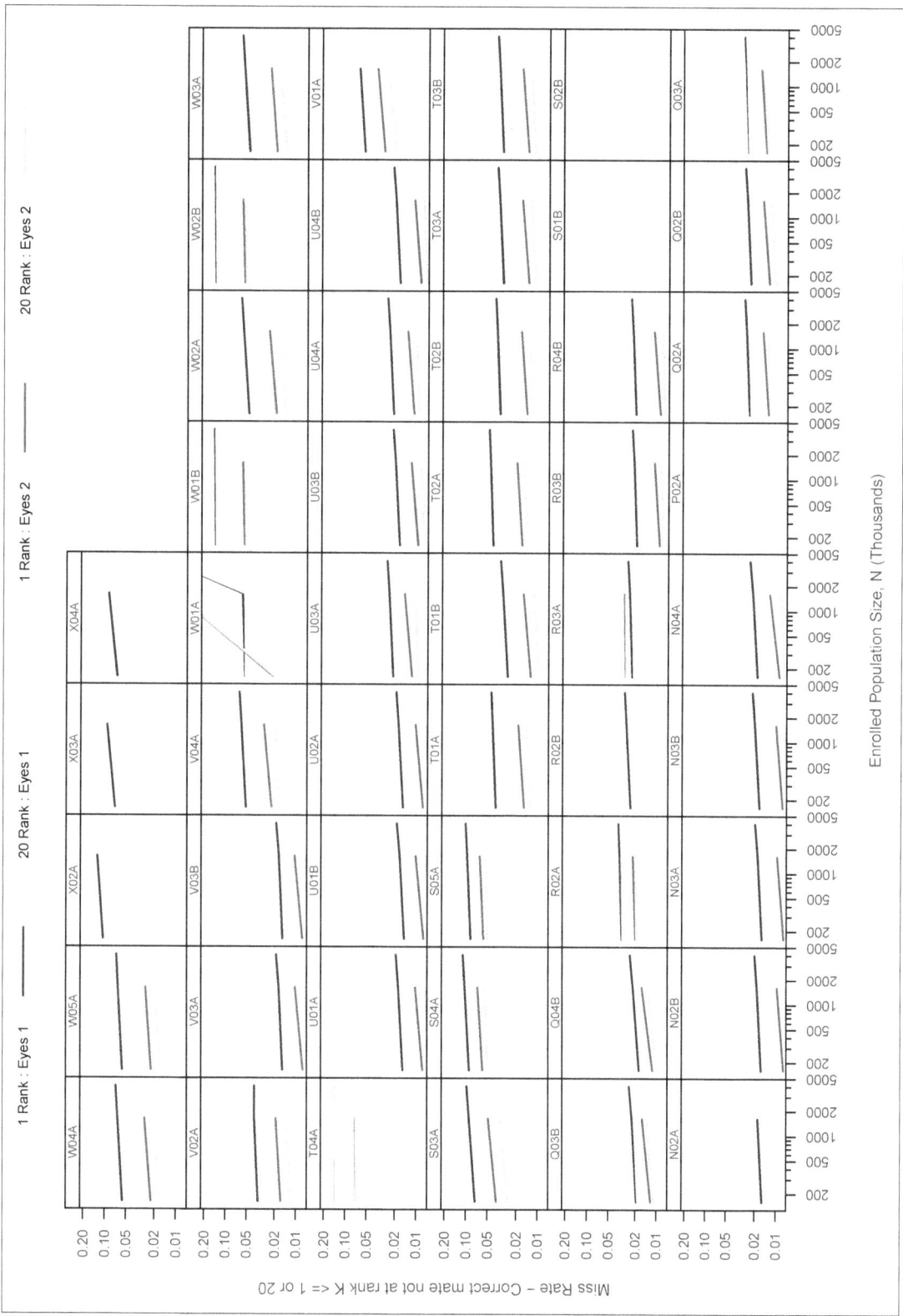

Figure 29: For group 0 SDKs received from February to June 2011, fraction of one-eye and two-eye searches for which a mate exists in the enrolled population but is not found at rank one (upper line), or at rank less than or equal to 20 (lower line) This statistic is plotted for enrolled populations of size $N = 20000, 160000, 1600000,$ and (one-eye only) 4000000 The statistic is estimated over the one-eye mated searches from the large sets S_{xb}, except for $N = 20000$ when S_{xa} are used instead For two-eye searching, the implementation may internally use one or both of the images Eye labels (L, R) were not provided

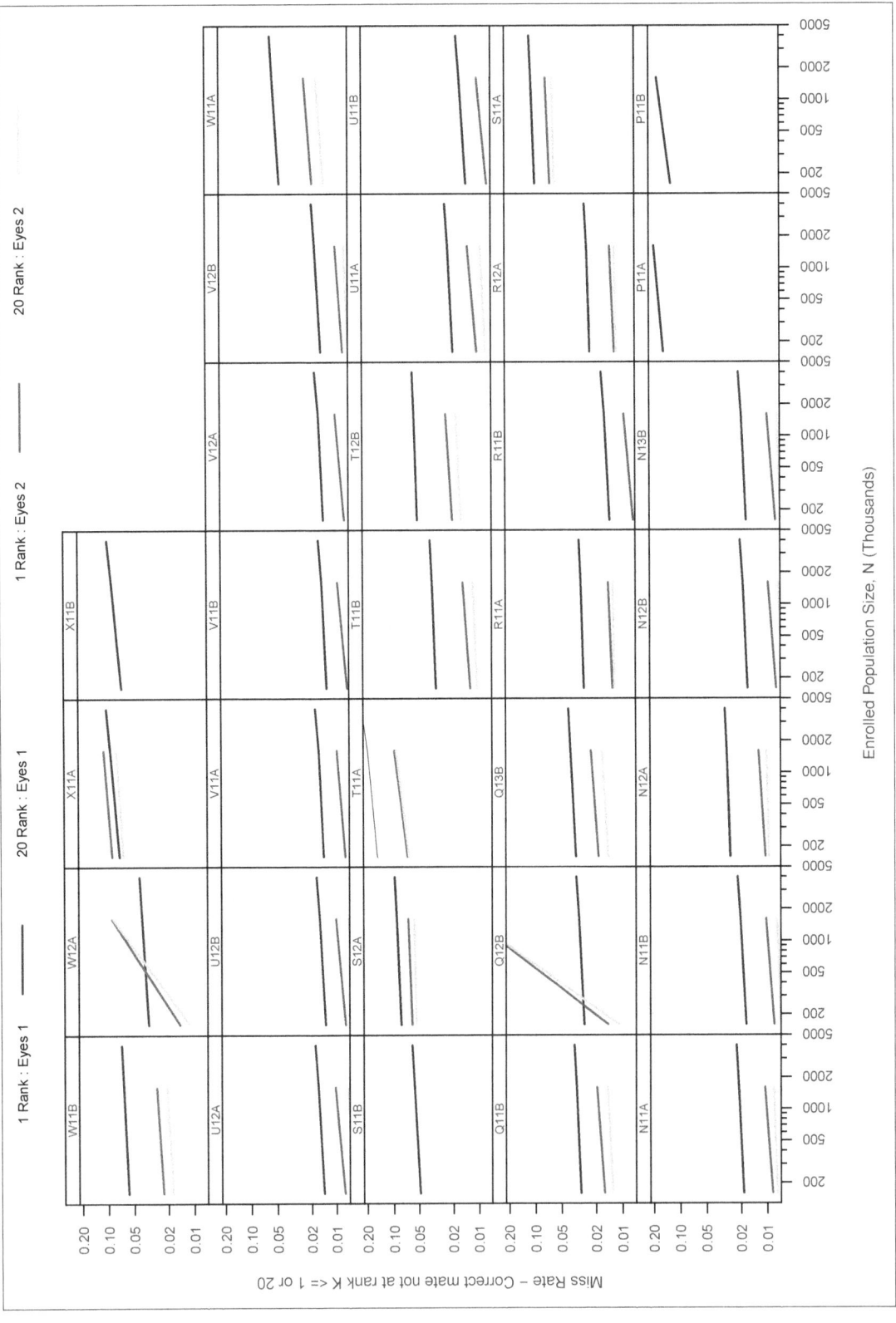

Figure 30: For group 1 SDKs received in August 2011, fraction of one-eye and two-eye searches for which a mate exists in the enrolled population but is not found at rank one (upper line), or at rank less than or equal to 20 (lower line) This statistic is plotted for enrolled populations of size N = 20000, 160000, 1600000, and (one-eye only) 4000000 The statistic is estimated over the one-eye mated searches from the large sets S_{xb}, except for $N = 20000$ when S_{xa} are used instead For two-eye searching, the implementation may internally use one or both of the images Eye labels (L, R) were not provided

8   Causes of failure

This section quantifies the effect of compression, iris size and iris dilation on recognition accuracy. In addition, it relates image quality to accuracy. The reader might first look at the IREX III FAILURE ANALYSIS supplement[21] which more completely details various image related effects implicated in recognition misses.

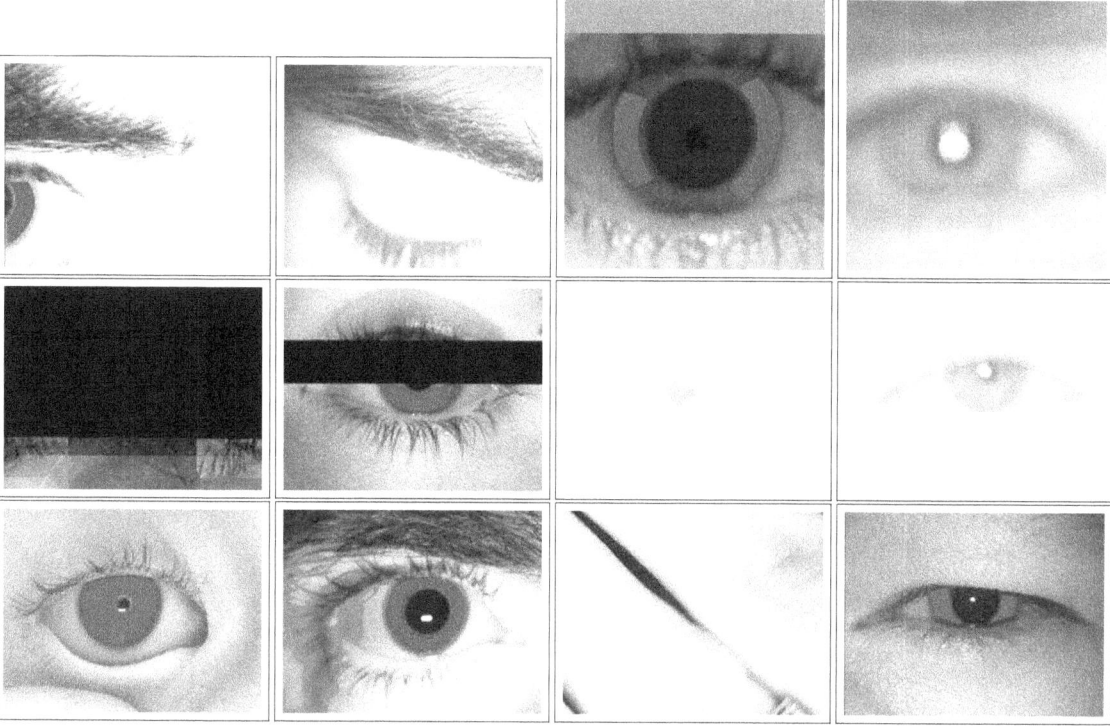

Figure 31:   Images that, for specific algorithms, produce aberrantly low dissimilarity nonmate scores. The red covering is applied by NIST to de-identify the iris. Some of these images, for example the last one, give such outcomes with more than one other image. These images would not give false positives in applications where the threshold is set lower than the minimums depicted in section 6.9.

8.1   Causes of false positives

By executing 311,427 nonmate searches against an enrolled population of N = 3,904,239 single eyes (see section 3.1), IREX III allows inspection of the lowest scoring nonmate image pairs from 1216 billion comparisons. The following are the authors' qualitative descriptions of those images, i.e. those for which dissimilarity is below threshold, τ, with $\text{FPIR}(\tau) < 5\ 10^{-6}$, and implied false match rate [22] $\text{FMR}(\tau) < 1.3\ 10^{-12}$. Generally the cause of failure can be categorized as follows

▷ **Defective images**: Figure 31 shows twelve images that result in low dissimilarity nonmate candidates; the other image of the pair is not shown. These images should be detected by image quality assessment software or suitably trained human operators.

▷ **Biological similarity**: Given that 1216 billion comparisons are conducted, some irides with biological similarity are expected, even with the threshold set so stringently. This is the case for most algorithms where the cause of the low-score cannot be attributed to a defective image.

---

[21]This document is linked from http://iris.nist.gov/irex

[22]While an FMR value of one part in a trillion is low, false matches will be expected unless the product of the enrolled population size and the number of search transactions is sufficiently small. For example, de-duplication of the 245 million residents of Indonesia, would involve $P(P-1)/2) = 6\ 10^{16}$ comparisons necessitating thresholds even lower than those reported in IREX III.

FNIR = FALSE NEGATIVE IDENT. RATE	N = NEUROTECHNOLOGY	P = SMU	Q = IRITECH	R = COGENT	S = SMARTSENSORS	T = CAMBRIDGE
FPIR = FALSE POSITIVE IDENT. RATE	U = L1	V = MORPHO	W = IRISID	X = CROSSMATCH	Y = KYNEN	

The sensitivities of the algorithms to various imaging defects are noted qualitatively below. These qualitative results are of primary use to algorithm developers who should a) attempt to replicate the result via synthetic manipulation of real images, and b) feel free to contact NIST to discuss.

▷ **N11A, N11B, N12A, N12B, N13B**: Saturated (white) images; images where the frame is incomplete (black rectangle across full width of image); dilated pupils in 330x330 compressed images. eye closed, with white eyelid. N13B seems less affected by saturated images.

▷ **Q11B, Q12B, Q13B**: Images for which iris is so poorly centered that it is cropped by one or two of the four edges of image.

▷ **R11A, R12A**: Motion blurred images; images where the frame is incomplete (black rectangle across full width of image) - which in some cases matches eyes where pupil is black, rectangular, and extends to the iris-sclera boundary; images with reflections from external environment.

▷ **R11B**: Images where the frame is incomplete (black rectangle across full width of image); dilated pupils in 330x330 compressed images.

▷ **S11A, S12A**: Occlusion from upper eyelid (into the pupil region).

▷ **S11B**: Presence of highly curved eyelids and fingers in the image; blurred images; images with reflections from external environment. constricted pupils.

▷ **T11A, T11B, T12B**: 330x330 compressed images.

▷ **U11A, U12A**: Quantized greylevel images; patterned contact lens.

▷ **U12B**: 330x330 compressed images; quantized greylevel images; patterned contact lens.

▷ **V11A, V12A**: 330x330 compressed images; highly quantized greylevel images.

▷ **V12B**: One enrolled image with low exposed iris area gives low scores with multiple others; one enrolled image with two fingers occluding iris gives low scores with two others; defocus blur; patterned contact lens.

▷ **W11A, W12B**: Dilated pupils; large radius irides; defocus blur.

▷ **W12A**: Patterned contact lens; quantized greylevel images.

▷ **X11A**: 330x330 compressed images; occlusion from upper eyelid (into the pupil region); images with reflections from external environment.

8.2   Effect of compression

Sample compression is important in all biometrics applications because samples are frequently transmitted across bandwidth-limited communications channels, or are stored in media of finite size (e.g. an e-Passport).

In IREX III all images were encoded using ISO/IEC 10918 JPEG compression. The filesizes and bit rate statistics are shown in Figure 32. The distributions are bimodal reflecting different parameterizations of the compressor. The first peak, BPP≤ 0.5, corresponds to a set of images of size 330x330 that are very heavily compressed (JPEG quality = 30) and exhibit tiling artifacts. Examples are shown in Figures 2(b) and 3(a). These images are excluded from much of the analysis in this report because, among all the images used here, they are *clearly* not representative of how a day-forward iris recognition application would be implemented. Indeed these typify why the vanilla JPEG algorithm is prohibited in formal iris standards because: accuracy can be degraded by use of too much compression; targeting specific filesizes requires iterative adjustment of the quality parameter of the JPEG algorithm, and because the ISO/IEC 19794-6:2011 and ANSI/NIST ITL 1-2011 standards include specific dedicated alternatives for transmission of minimum filesize irides. The

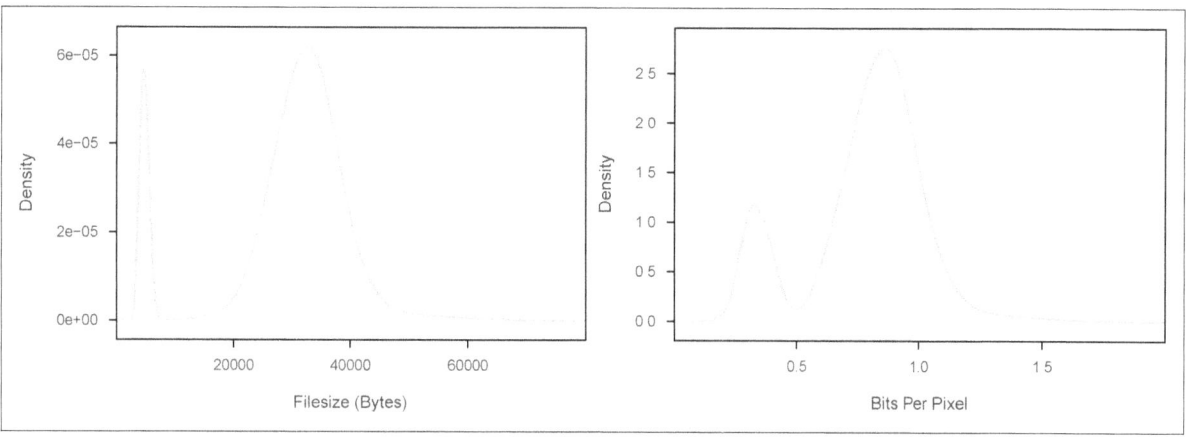

Figure 32: Filesizes and equivalent bit rates for the 6.1 million images of the OPS population.

second peak Figure 32 corresponds to application of JPEG QUALITY = 75; and these images are used to compute the main results of this report.

The empirical effect of image compression is presented in the heatmaps of Figure 33. The Figures use binned bits-per-pixel (BPP) values, the histogram of which is shown in Figure 32. For mate searches, the BPP values for the search image and enrolled mate image are used. For nonmate searches, the BPP values for the search image and its top nonmate candidate are used. Three heatmaps are generated: one for the mean dissimilarity score; one for the observed FNIR at FPIR = 0.0001 in a population of 1.6 million; and the third showing the number of image pairs in the bin. Analogous figures for all algorithms are presented in the IREX III APPENDICES linked from http://iris.nist.gov/irex.

The notable observations are as follows.

▷ Mate scores and FNIR values ascend as BPP values diverge from a best value of 1 bit per pixel. This is obviously true for small BPP values but the cause of degraded recognition at larger sizes BPP$\geq$ 1.2 is the presence of images like that shown in Figure 3(b). These images include dithered and quantized regions that do not compress well.

▷ The very low bit rate bins $0 \leq$ BPP ≤ 0.375 and $0.375 \leq$ BPP ≤ 0.5 are occupied exclusively by the 330x330 pixel images shown in 2(b) and 3(a). The error rates vary for these images substantially.

▷ Use of JPEG compression is prohibited in formal standards (ISO/IEC 19794-6 and ANSI/NIST ITL 1-2011 because elevated false positive and false negative errors result when too much compression is applied.

FNIR = FALSE NEGATIVE IDENT. RATE	N = NEUROTECHNOLOGY	P = SMU	Q = IRITECH	R = COGENT	S = SMARTSENSORS	T = CAMBRIDGE
FPIR = FALSE POSITIVE IDENT. RATE	U = L1	V = MORPHO	W = IRISID	X = CROSSMATCH	Y = KYNEN	

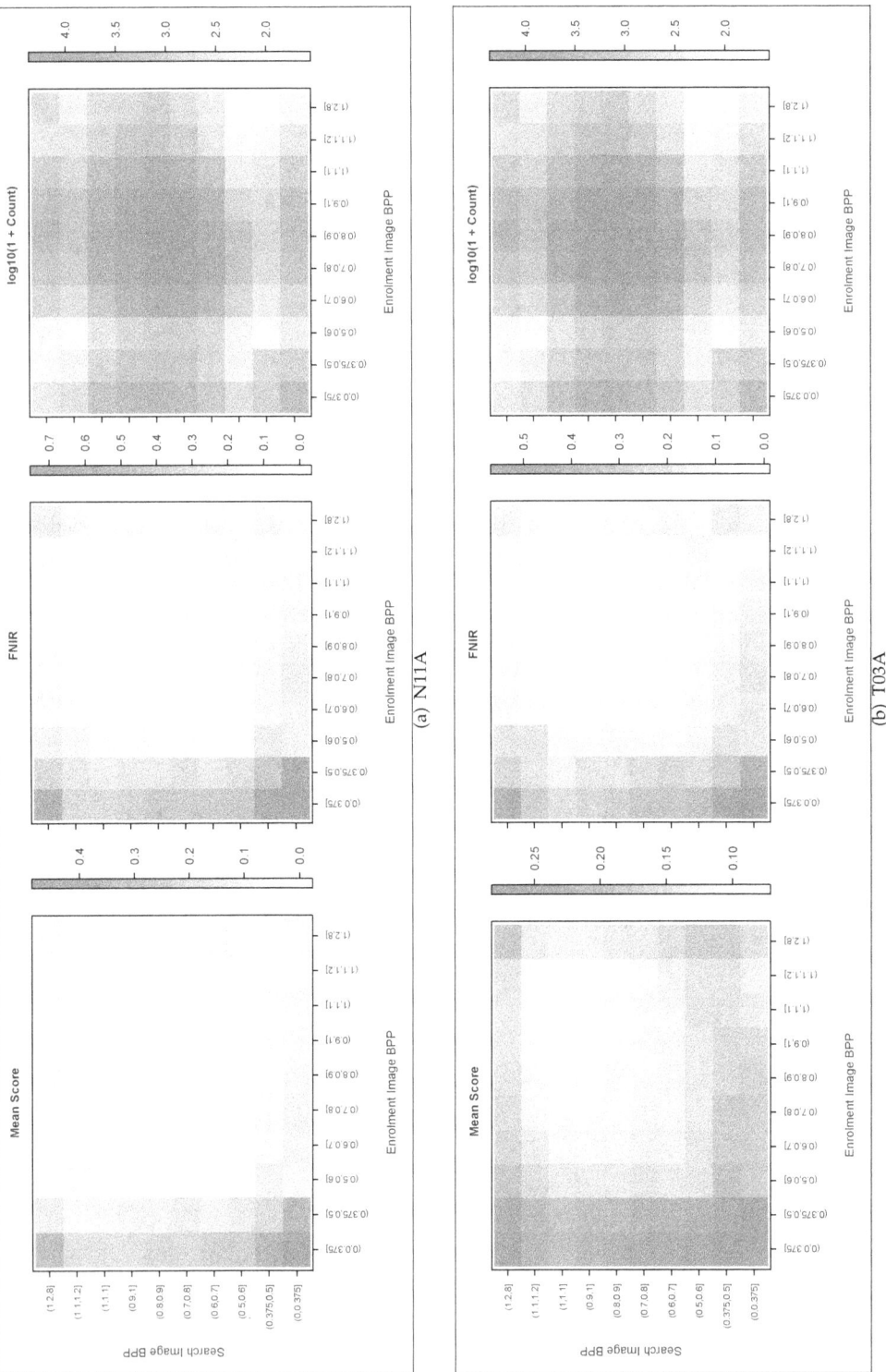

Figure 33: **Effect of compression on false negatives** for implementations N11A and T03A executing searches in an enrolled population of N = 16 million The axes are labeled by binned values of the number of bits per pixel (BPP) implied by the JPEG filesize and the image dimensions The lowest BPP bins are populated largely by the overly compressed 330x330 images The three heatmaps give, respectively, the score, FNIR at FPIR = 0.0001 and the logarithm of the count of images in each bin (3 corresponds to 1000) For other algorithms, see analogous figures given in IREX III APPENDIX E

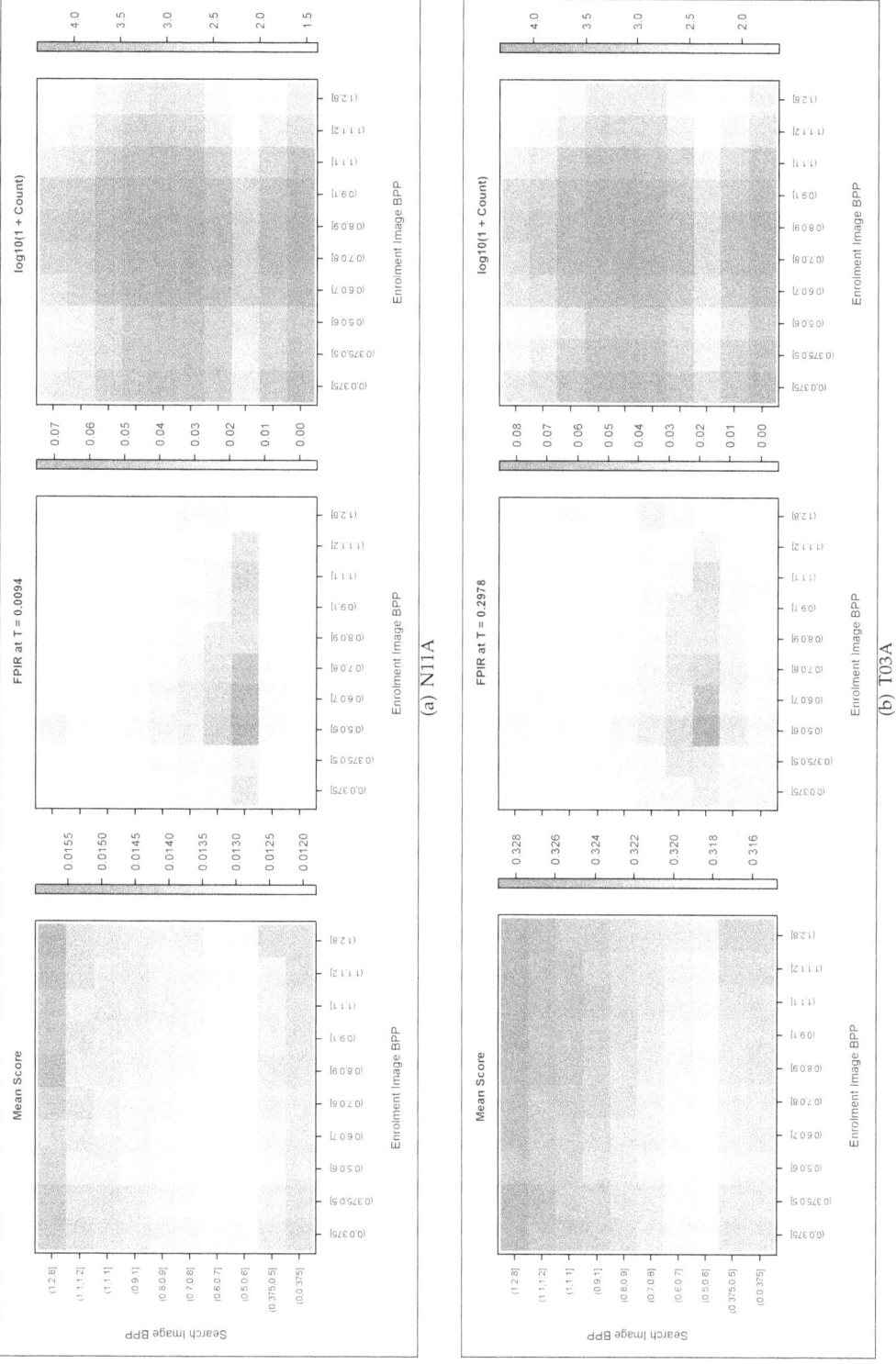

Figure 34: **Effect of compression on false positives** for implementations N11A and T03A executing searches in an enrolled population of N = 1.6 million. The axes are labeled by binned values of the number of bits per pixel (BPP) implied by the JPEG filesize and the image dimensions. The lowest BPP bins are populated largely by the overly compressed 330x330 images. The threshold is set to give FPIR = 0.001 globally. For other algorithms, see analogous figures given in IREX III APPENDIX E.

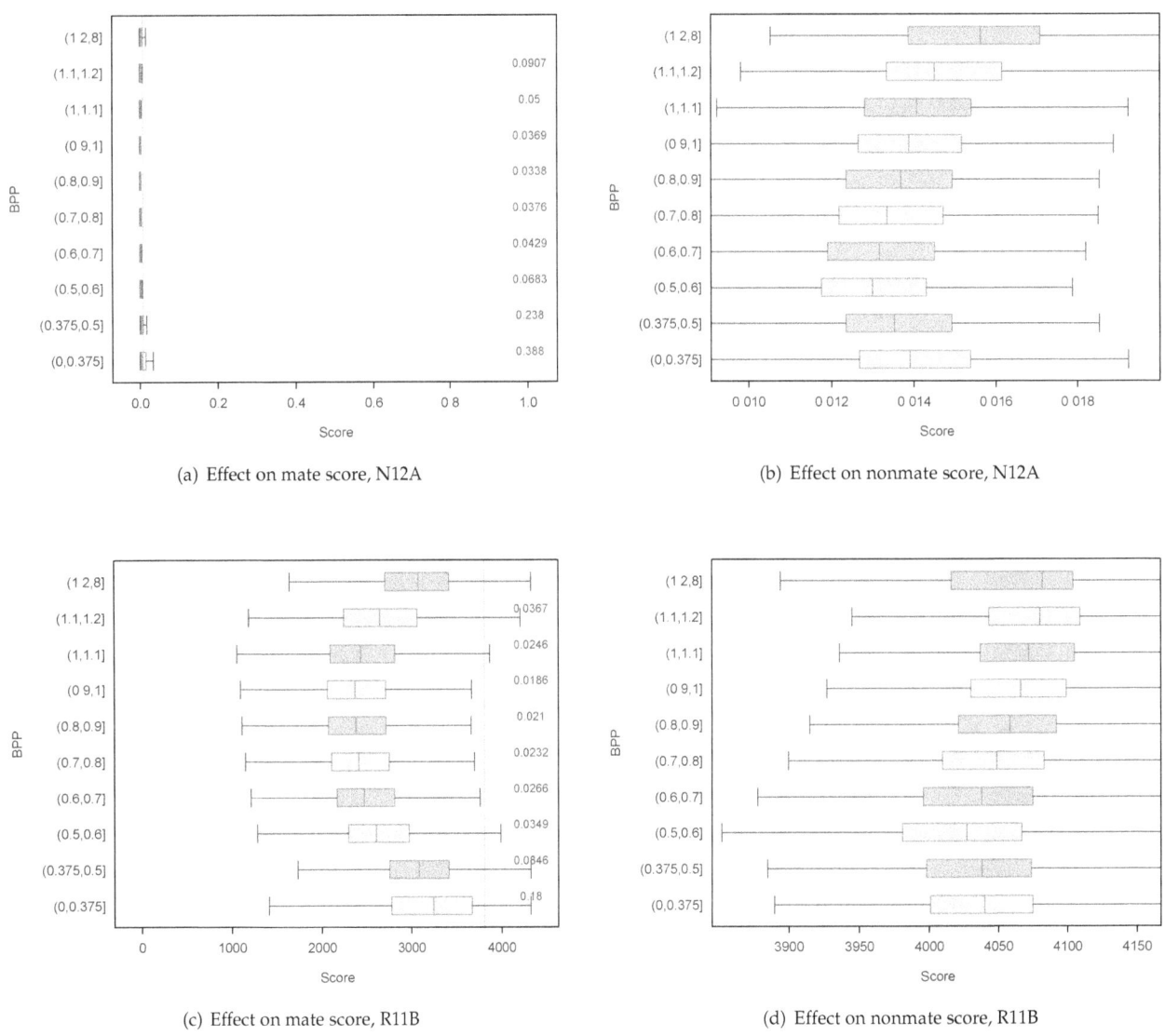

(a) Effect on mate score, N12A

(b) Effect on nonmate score, N12A

(c) Effect on mate score, R11B

(d) Effect on nonmate score, R11B

Figure 35: **Effect of compression on dissimilarity score**: For two implementations executing searches in an enrolled population of N = 1.6 million, the vertical axis gives the number of bits per pixel (BPP) implied by the JPEG filesize and the image dimensions. For any given comparison, the minimum of enrolled candidate and search image BPP values is used. The lowest two BPP bins are populated largely by the overly compressed 330x330 images. The boxplot shows the effect on the score, with the FPIR value annotated at right. For other algorithms, see analogous figures given in the Appendix E.

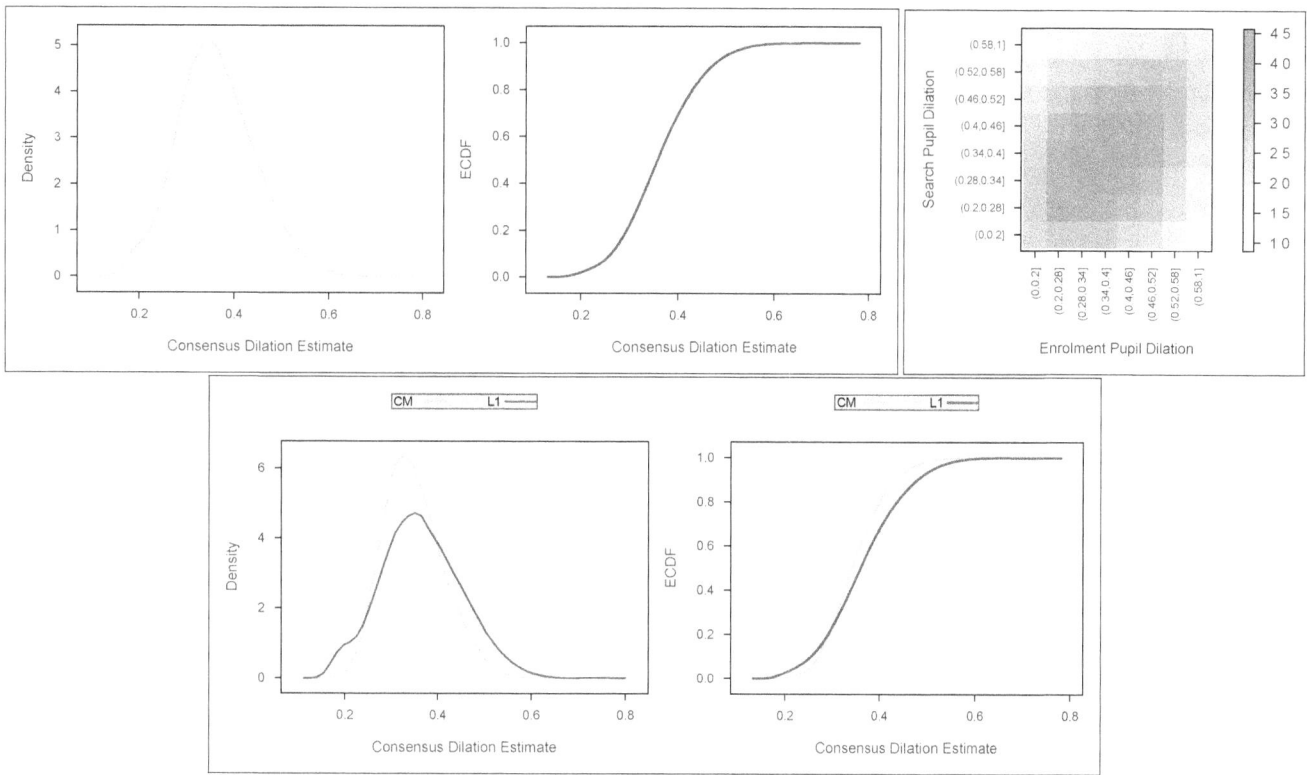

Figure 36: On the **left** are empirical density and distribution functions of pupil radius to iris radius ratio. These are consensus estimates made by taking the median of an image's radius estimates from all implementations. To the **right** is a heatmap, $\log_{10}(1 + N_{ij})$, of the number of mated image pairs, N_{ij}, whose pupil-iris ratio estimates fall in (unevenly spaced) bins i and j. There are very few pairs for which dilations differ maximally. **Below** is the empirical density and distribution functions for the subset of images known to have been collected with the two most common camera families, L1 for L1 Identity Systems and CM for Crossmatch Technologies. The cameras operate in different modes. The L1 PIER and HIIDE devices are handheld at distances of at least 18 centimeters; the CM SEEK devices use a binocular-like capture. The gold lines give the bin delimeters used in subsequent figures. These are 0, 0.2, 0.28, 0.34, 0.40, 0.46, 0.52, 0.58, 1.

8.3   Effect of pupil dilation and constriction

Iris recognition algorithms are required to compensate for changes in a subject's pupil dilation between initial enrollment and subseqeuent search. In addition, neither the low iris area inherent in a dilated pupil, nor the large area associated with a constricted pupil, should negate the ability to discriminate a particular search iris with all those enrolled.

8.3.1   Background

One challenge arises, at least for implmentations that use a radial and circumferential sampling schems to establish a polar representation of the iris texture, because a dilated pupil implies a radially narrowed iris texture and the need to sample this without (typically) better optical resolution. Another challenge is that the deformation of the iris, as a three dimensional anatomical object, is not known to present a linear two dimensional projection to a camera. However, as deformation is locally linear, the problem is to know over what range is the linearity model sufficient. The original Daugman patent[23] claims a technique "to extract and encode an iris signature that remains the same over a wide range of pupillary dilations". It achieved this via the linear treatment implied by the doubly dimensionless representation of the iriscode.

Recent one-to-one verification studies have documented an relationship between dilation accuracy. Hollingsworth[19]

---

[23] United States Patent 5291560, *Biometric personal identification system based on iris analysis*, filed July 15 1991, published March 1, 1994

concluded that mean mate Hamming Distances (from a non-commercial implementation of the Daugman algorthm) increased by 0.06 for dilated eyes, and by 0.08 for differently dilated eyes. The study also concluded that the mean of the distribution for nonmate pairs of highly dilated eyes ($R_p/R_i > 0.6$) was 0.02 lower than for less dilated pairs. Howver, there was no discernible effect on the nonmate dirtsribution for differences in dilation. The study quantified difference in dilation in terms of iris and pupil radii as

$$\Delta D = D^{(2)} - D^{(1)} = \frac{R_p^{(2)}}{R_i^{(2)}} - \frac{R_p^{(1)}}{R_i^{(1)}} \tag{14}$$

where the superscripts indicate the first (enrollment) image, and the second (search) image.

The IREX I evaluation[12] also considered the dilation issue, for commerical one-to-one verification algorithms applied to uncompressed and compressed images. It quantified the effect in terms of false match and non-match rates, FMR and FNMR, and image-specific variants of those. The study measured an increase in both false positive and negative outcomes for high dilation, and a dependence on change in dilation, $1 - \alpha$, where dilation consistency, α, was defined as

$$\alpha = \frac{1 - \max(D^{(1)}, D^{(2)})}{1 - \min(D^{(1)}, D^{(2)})} = \left(\frac{R_i^{(2)}}{R_i^{(1)}} \right) \left(\frac{R_p^{(1)} - R_i^{(1)}}{R_p^{(2)} - R_i^{(2)}} \right) \tag{15}$$

where D is the pupil iris ratio, and the first parenthesized ratio, which compensates for difference in camera magnification, is cleanly separated from the second which is the ratio of the smaller to the larger radial iris thickness. The superscripts indicate the first (enrollment) image, and the second (search) image. The result was that mate scores increase with difference in dilation, but that nonmate scores do not. The IREX I study was also able to consider the combined effects of dilation and occlusion, the result being that both factors give higher false negatives but with relative weights specific to the particular recognition algorithm.

IREX II - IQCE was conducted to support development of iris quality factors for the ISO/IEC 29794-6 standard. The draft of that standard includes dilation consistency as a measure of quality for pairs of images. It is defined as

$$\alpha = \frac{\min(D^{(1)}, D^{(2)})}{\max(D^{(1)}, D^{(2)})} \tag{16}$$

where D is again the pupil iris ratio, and $0 \leq \alpha \leq 1$ with higher values indicating dilation in two images is similar.

The IREX II work found matcher-dependent relationships of dilation (estimates taken directly from quality assessment algorithms) and both false negative and false postive rates (see [30], Figure 38a).

8.3.2   Results

The data used in IREX III has dilation statistics shown in Figure 36 as estimated over the 1.6 million enrollment and 0.5 million search images. These statistics differ considerably from those reported over the nearly 21708 IrisID IrisAccess 4000 images collected and studied at NOTRE DAME(ND)[26]. For the IREX III data the mean pupil iris ratio is 0.36[24] while for for the ND data the mean (by visual inspection) is 0.42. Further, only a few percent of the ND images have dilation below the IREX III quartile of 0.31, and fully 50% of it lies above the IREX III third quartile of 0.42. This difference may be demographic, ethnologic and environmental causes: the ND images were collected in a laboratory environment, and while IREX III corpus in various detainee enrollment facilities some of which were close to outdoor sunshine. The dilation differences were originally noted in the orginal IREX I report[12] for the OPS and ICE images which hail, respectively, from the same two sources.

For IREX III, the empirical effect of pupillary dilation and constriction is presented for the images of the large set $\mathcal{S}_{1b}$ as searched against an enrollment set of N = 1,600,000. Pupil and iris radii are reported by each implementation. Unless stated otherwise each image is assigned a consensus radius computed as the median over all algorithms' estimates. The

---

[24] Minimum is 0.1316, first quartile is 0.3063, median is 0.3571, mean is 0.3621, third quartile is 0.4151 and maximum is 0.781

FNIR = FALSE NEGATIVE IDENT. RATE	N = NEUROTECHNOLOGY	P = SMU	Q = IRITECH	R = COGENT	S = SMARTSENSORS	T = CAMBRIDGE
FPIR = FALSE POSITIVE IDENT. RATE	U = L1	V = MORPHO	W = IRISID	X = CROSSMATCH	Y = KYNEN	

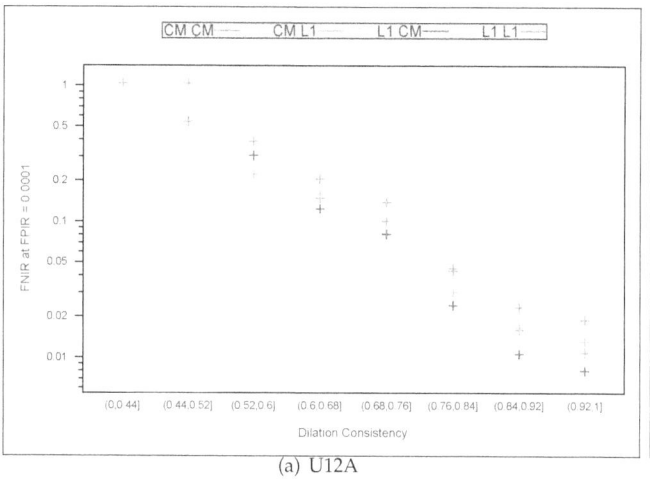
(a) U12A

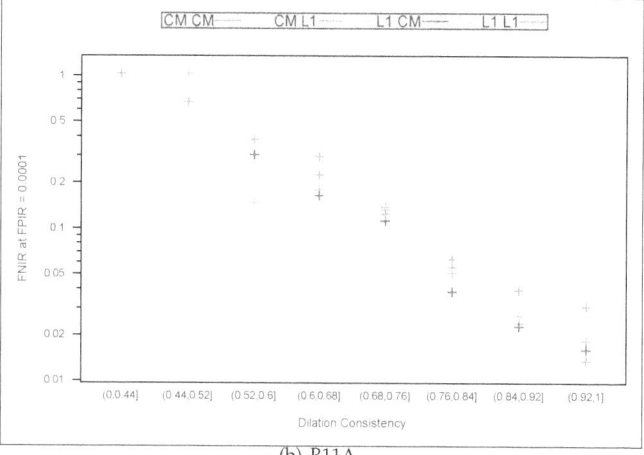
(b) R11A

Figure 37: **Effect of dilation and constriction**: For two algorithms the panels show, for paired camera families denoted CM and L1, mean FNIR for binned dilation consistency values of mated pairs of images per equation 15. See Figure 36 for estimated distributions of the input dilations. The threshold gives FPIR = 0.0001 globally. Searches not producing the correct candidate were assigned a mate score equal to the highest observed mate score. The searches were from set $\mathcal{S}_{1b}$. The enrolled population was $N = 1,600,000$. Larger figures for each algorithm appear in the IREX III APPENDICES.

notable results regarding false negative identification rate FNIR are as follows.

▷ **Pupils in both search and mate image are dilated**: Figure 39 shows, for all algorithms including the most accurate ones, FNIR is elevated when both the search image and its mate are dilated with pupil-iris ratio above 0.6. Dilation may present a problem for two reasons: the reduced iris texture area limits information content, and the need to search for large pupils.

▷ **Pupils in both search and mate image are constricted**: All algorithms give elevated FNIR when the pupil is constricted with pupil-iris radius ratio below 0.2. This effect is larger than for dilation. Constriction may be problematic becuase algorithms are not configured to search for small pupils.

▷ **Pupils in search and mate images are differently dilated**: The most severe FNIR increase related to pupil dilation and constriction arises for all algorithms when the pupils are differently dilated. This appears in red in the off-diagonal elements in Figure 39. Figure 38 shows FNIR as a function of the constriction-dilation consistency measure of equation 15. This essentially integrates the heatmap errors parallel to the diagonal. All algorithms show a characteristic increase in false negatives with divergence in the two pupil-iris ratios. Specifically, the logarithm of FNIR increases about linearly with the consistency value α in eq. 15. This corresponds to an exponential dependence of FNIR on the ratio of the radial iris thicknesses (measured in the object plane, not the digital image). The large increases in FNIR for $\alpha < 0.8$ would contra-indicate enrollment of atypically dilated or constricted pupils. For example, a first sample with median dilation ($D_1 = 0.33$), this limit implies the second iris should have dilation $0.18 \leq D_2 \leq 0.47$. For a first sample $D_1 = 0.36$ the second should have $0.2 \leq D_2 \leq 0.49$.

▷ The plotting of α from equation 15 is preferred over 16 because of its physical meaning, and because the latter gives a sigmoid-like response FNIR(α) vs. the linear one reported in Figure 38. This issue is pertinent to the draft ISO/IEC29794-6 iris image quality standard.

▷ None of the algorithms appear markedly more tolerant high pupillary dilation and constriction than others.

In addition, dilation and constriction have an effect on false positive identification rate FPIR as detailed below. Note that FPIR is estimated over nonmate searches so all returned candidates are from different irides.

▷ **Pupils in both search and nonmate images are constricted**: The bottom leftmost cell in the panels of Figure 42 shows for some Q, V, and R algorithms, that a constricted pupil tends to return candidate images also with constricted pupils. For the X and early V algorithms, the nonmate candidates can have larger, more normal, dilation values i.e. it constriction in the search image alone tends to produce low scoring nonmate candidates.

▷ **Pupils in both search and nonmate images are dilated**: As the other end of the dilation range, Figure 41 shows that some R and all Q algorithms tend to return nonmate candidates also with high dilation, and these candidates have low dissimilarity values. Many other algorithms show some tendency to do this too, W, V, T, S. Two factors may be implicated in this effect: Iris area lost to eyelid and eyelash occlusion, and reduced iris area generally. Compensation for these has been discussed in the literature[22] and associated with dilation[19].

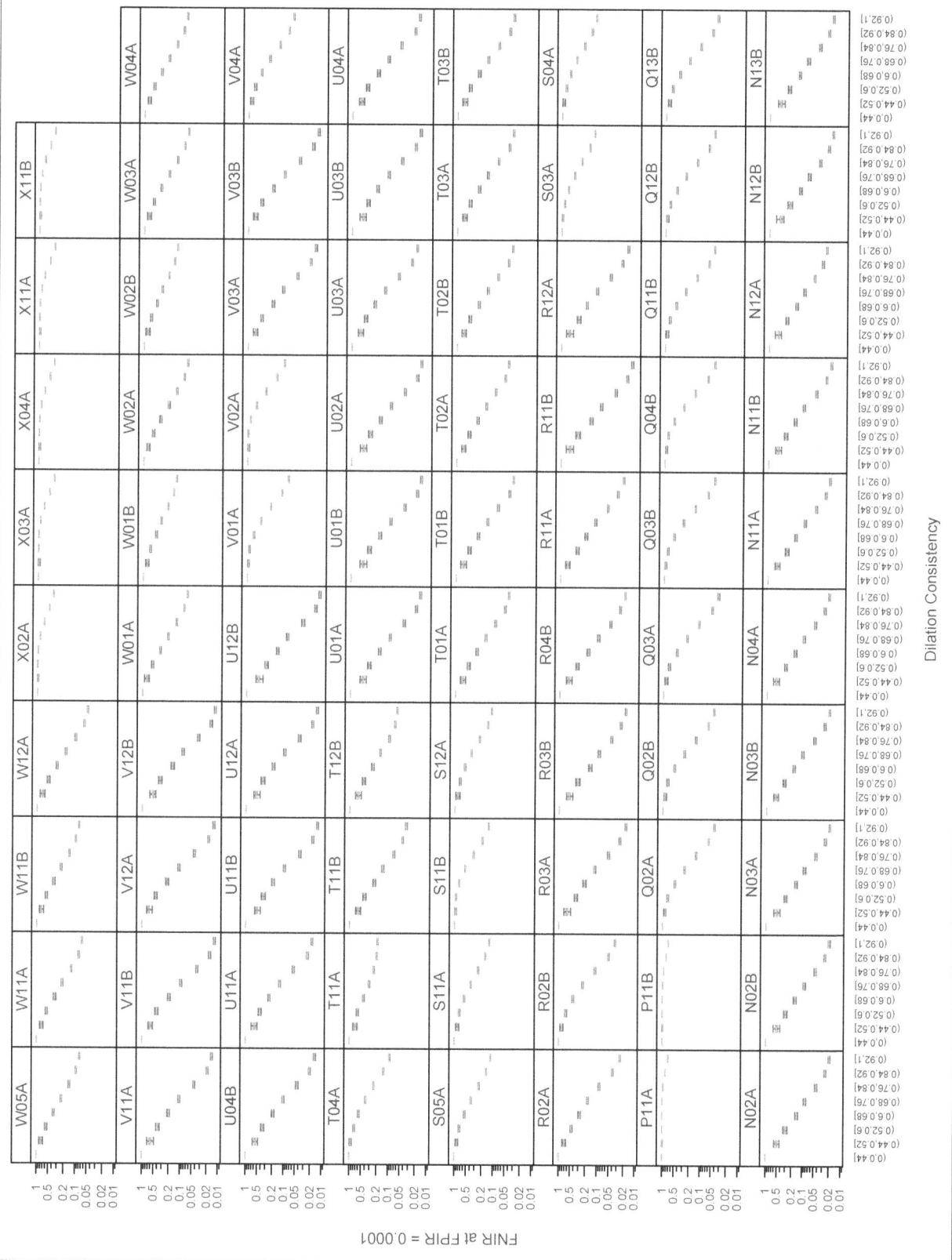

Figure 38: **Effect of dilation and constriction**: Each panel shows bootstrap estimates of FNIR for binned dilation consistency values of mated pairs of images per equation 15 for radii computed as the median of the radii reported by all the implementations See Figure 36 for estimated distributions of D The threshold gives FPIR = 0 0001 globally Searches not producing the correct candidate were assigned a mate score equal to the highest observed mate score S_{1b} The enrolled population was $N = 1,600,000$ The number of mated comparisons are, from left to right, 6, 100, 706, 2700, 8941, 25196, 61348, 113826 and thus statistical significance is poor on the far left side Larger figures for each algorithm appear in the IREX III APPENDICES

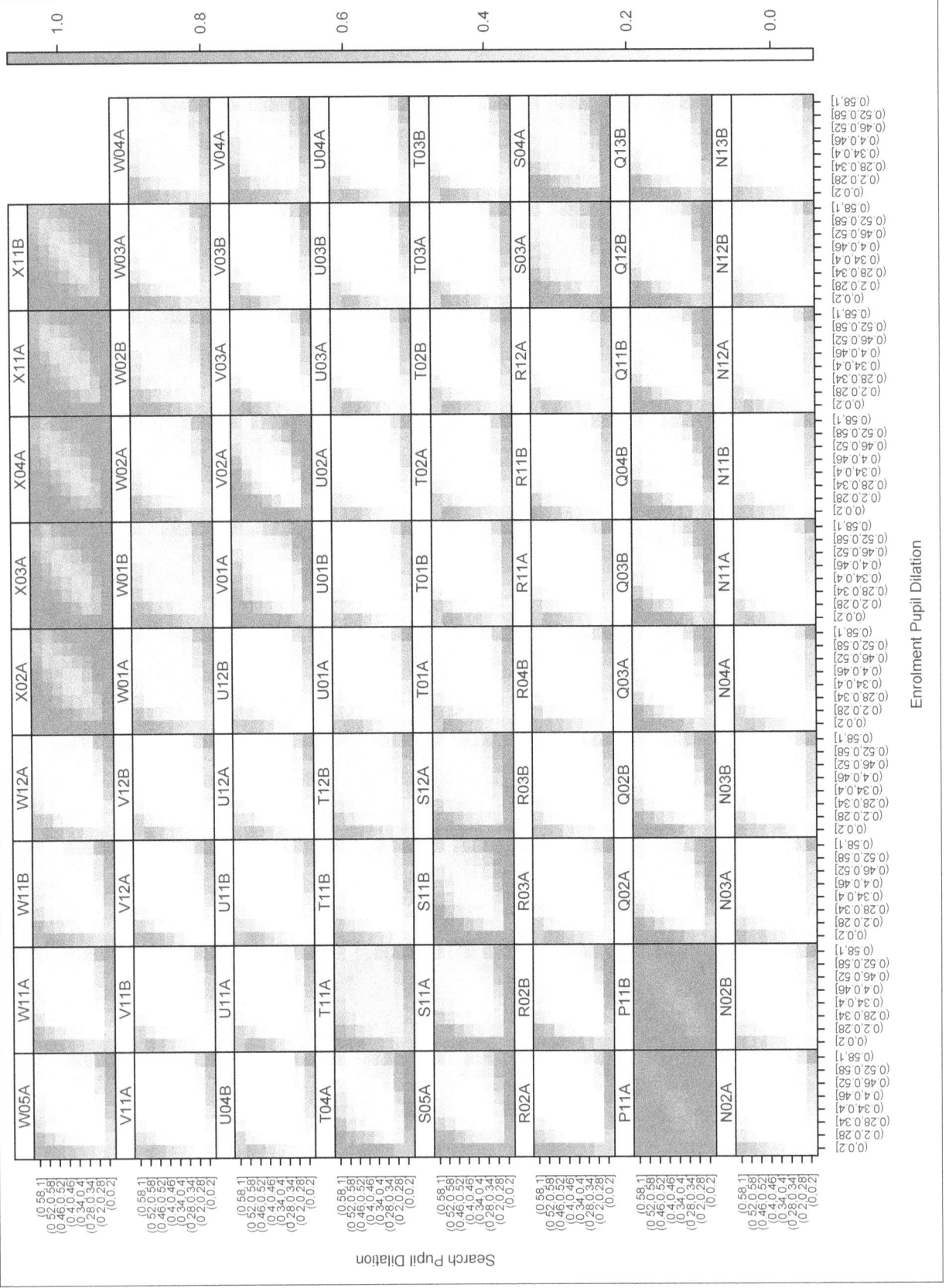

Figure 39: **Effect of dilation and constriction:** Each panel shows FNIR for binned dilation values of the enrollment and search images. The bins have unequal width. The dilation value is the pupil-iris radius ratio R_p/R_i, where, for any given image, R is the median of the radii reported by all the implementations. See Figure 36 for estimated distributions of pupil-iris ratio and for mated pairs. The threshold was set to give FPIR = 0.0001. For this analysis, searches that did not produce the correct candidate were assigned a mate score equal to the highest observed mate score. The mate searches were those of the large search set S_{1b}. The enrolled population size was $N = 1,600,000$. Larger figures for each algorithm are given in the IREX III APPENDICES

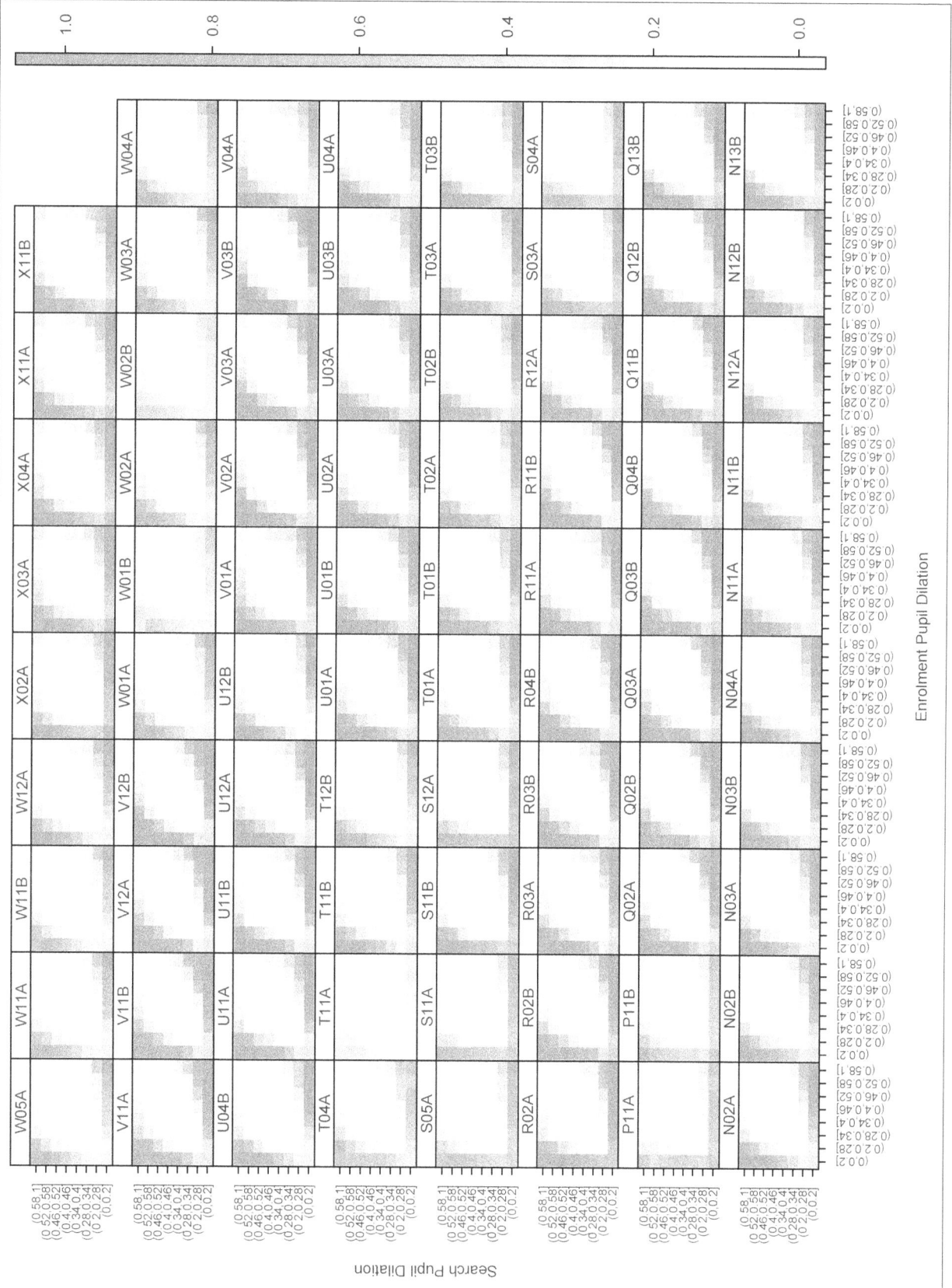

Figure 40: **Effect of dilation and constriction:** Each panel shows, for binned dilation values, the proportion of normalized **nmate** dissimilarities that are above 0.9. Over the whole panel the expected value is 0.1; in any specific cell a departure from that value indicates a sensitivity to the pupil-iris radius ratio R_p/R_i. A value of 1.0 (in red) indicates that all of the dissimilarities are above the 90% percentile of the global mate score distribution. The dissimilarities are normalized via $0 \leq F(x) \leq 1$ where F is the empirical cumulative distribution function of all mate scores produced from the mate searches of the large search set S_{1b}. R is the median of the radii reported by all the implementations.

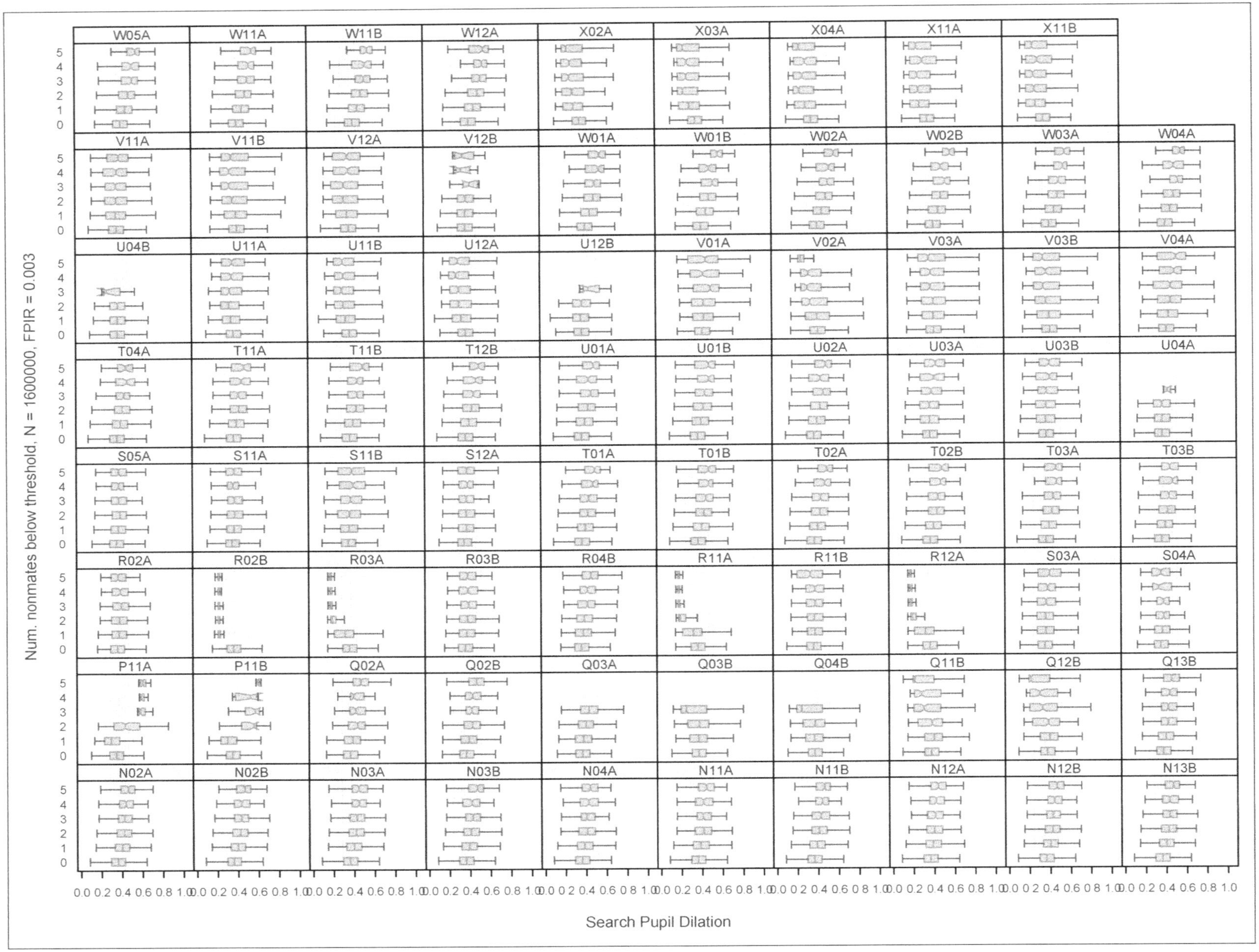

Figure 41: **Effect of dilation and constriction:** Each panel shows on the horizontal axis the distribution of the dilation estimate of the search image, i.e. pupil-iris radius ratio R_p / R_i. The yellow lines give the 10, 20 . . . 90 percentiles. R is the radius reported by the algorithm identified in the panel header. The vertical axis gives "selectivity" i.e. the number of nonmates on a candidate list below threshold. The threshold was set to give FPIR = 0.003 over all searches of the large search set S_{lb} for an enrolled population size of $N = 1,600,000$ and a candidate list length of 20. The overly compressed 330x330 images are excluded. Any given search can produce from 0 to 20 nonmates less than the threshold. Thus, the panels show dilations for searches that produce 0 . . . 5 false matches. A trend upwards and to the right indicates the algorithm gives more false matches when the pupil is dilated; upwards to the left indicates a sensitivity to pupil constriction.

| FNIR = FALSE NEGATIVE IDENT. RATE | N = NEUROTECHNOLOGY | P = SMU | Q = IRITECH | R = COGENT | S = SMARTSENSORS | T = CAMBRIDGE |
| FPIR = FALSE POSITIVE IDENT. RATE | U = L1 | V = MORPHO | W = IRISID | X = CROSSMATCH | Y = KYNEN | |

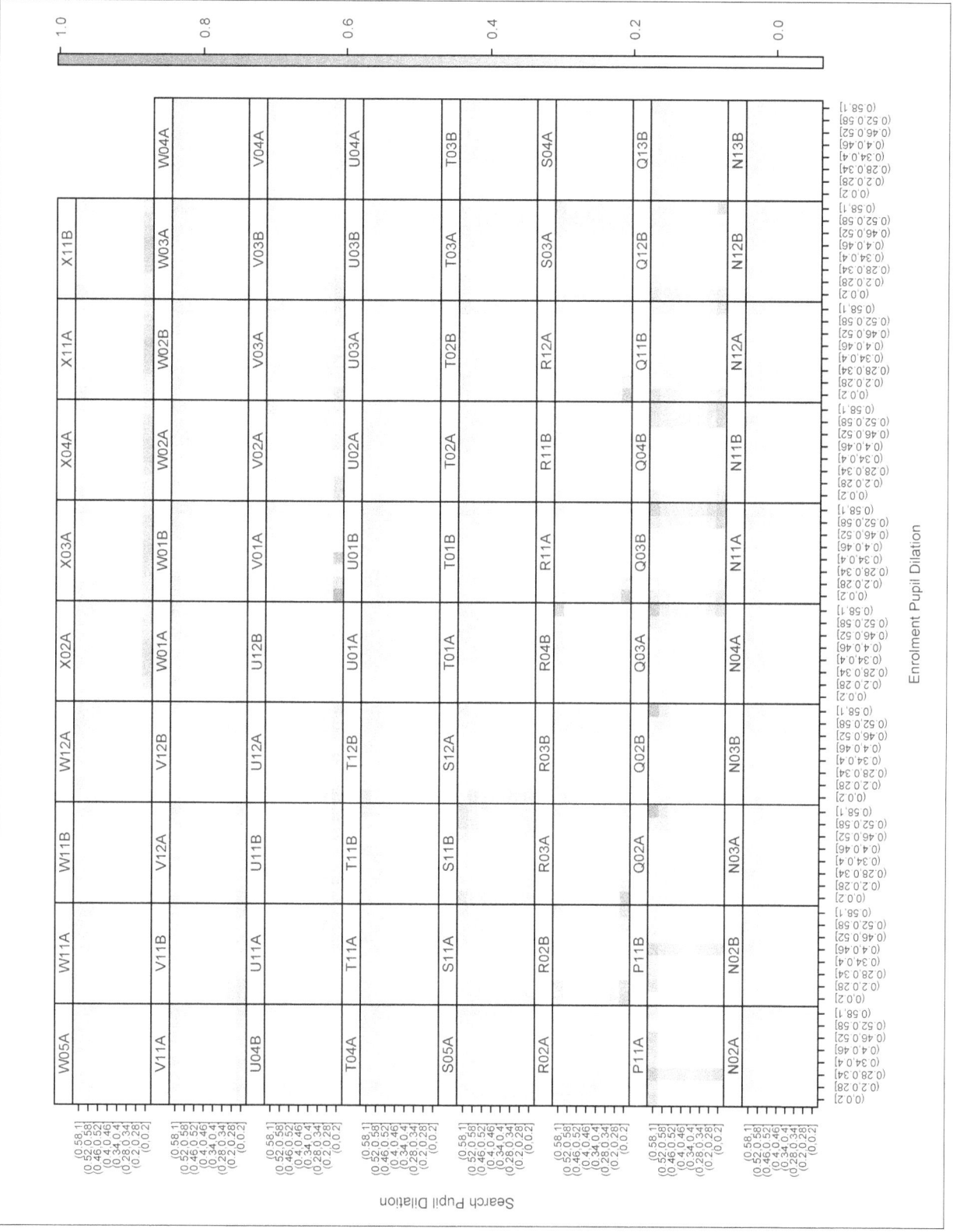

Figure 42: **Effect of dilation and constriction:** Each panel shows, for binned dilation values, the proportion of normalized **nonmate** dissimilarities that are below 0 1 Over the whole panel the mean is 0 1; in any specific cell a departure from that value indicates a sensitivity to the pupil-iris radius ratio R_p/R_i A value of 1 0 (in red) indicates that all of the dissimilarities are in the lower 10% of the global nonmate score distribution For any given image, radii are those reported by the algorithm identified in the panel header The dissimilarities are normalized via $0 \le F(x) \le 1$ where F is the empirical cumulative distribution function of all nonmate scores produced from the nonmate searches of the large search set S_{1b} The enrolled population size was $N = 1,600,000$ The overly compressed 330x330 images are excluded

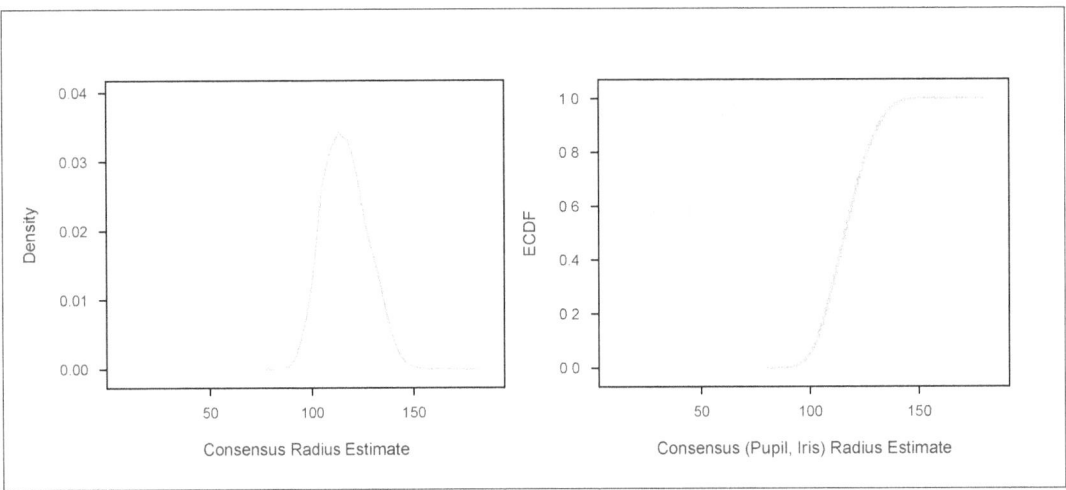

Figure 43: **Pupil and iris radii**: Empirical density and distribution function estimates for iris and pupil radii. For any given image, the radii are the median of the values reported by *all* tested algorithms; the median is resistant to poor estimates.

8.4 Radius

The IREX I study showed that algorithms fail when presented with images containing irides of unexpected radii. That is, algorithms had been developed around a de facto standard 640 × 480 image format in which iris radii from commercial cameras were targeted to be on the range $[80, 160]$. The IREX III results in this section show again that small and large irides are associated with higher (i.e. poor) dis-similarity scores and false negative outcomes.

Note that pupils and irides are rarely perfect circles - irregular shape is part of the challenge for iris recognition, and the subject of considerable research. The IREX III API did not include an encoding for complex boundary shapes, and consideration of such is not yet possible from the measured results.

8.4.1 Methods

The template generators used in IREX III report the radii of the pupil and the iris. While circular models of iris and pupil have long been deprecated, these radii are useful for analysis. Given K implementations reporting radii R_k for a given image, a consensus iris radius is formed as the median of the R_k $k = 1 \dots K$ estimates, without any attempt to weight the goodness of the algorithm. Consensus pupil radii are calculated similarly. The radius statistics are presented in Figure 43. The vertical lines define quantile bins A = $(52, 102]$ B = $(102, 106]$ C = $(106, 108]$ D = $(108, 111]$ E = $(111, 114]$ F = $(114, 116]$ G = $(116, 119]$ H = $(119, 122]$ I = $(122, 124]$ J = $(124, 128]$ K = $(128, 133]$, and L = $(133, 189]$ These are used in the analysis of the next section. The bins labeled A-L are quantiles of the observed iris radius distribution:

8.4.2 Results

Figure 44 shows false negative accuracy for the various enrollment and search sample iris radius ranges. For each cell in each panel, the statistic is FNIR(R_E, R_S) minus the mean FNIR over all radii. The occurrence of red color indicates overall high error rates from some algorithms. The distribution within each panel generally increased FNIR at low radii, and less frequently, at high radii too. Importantly, this may not be the result of low radius in and of itself, but rather that out-of-focus irides too far from the camera (i.e. beyond the depth of field) may result in low radius.

In any case, the observed iris size dependence is not just a product of the template generator.

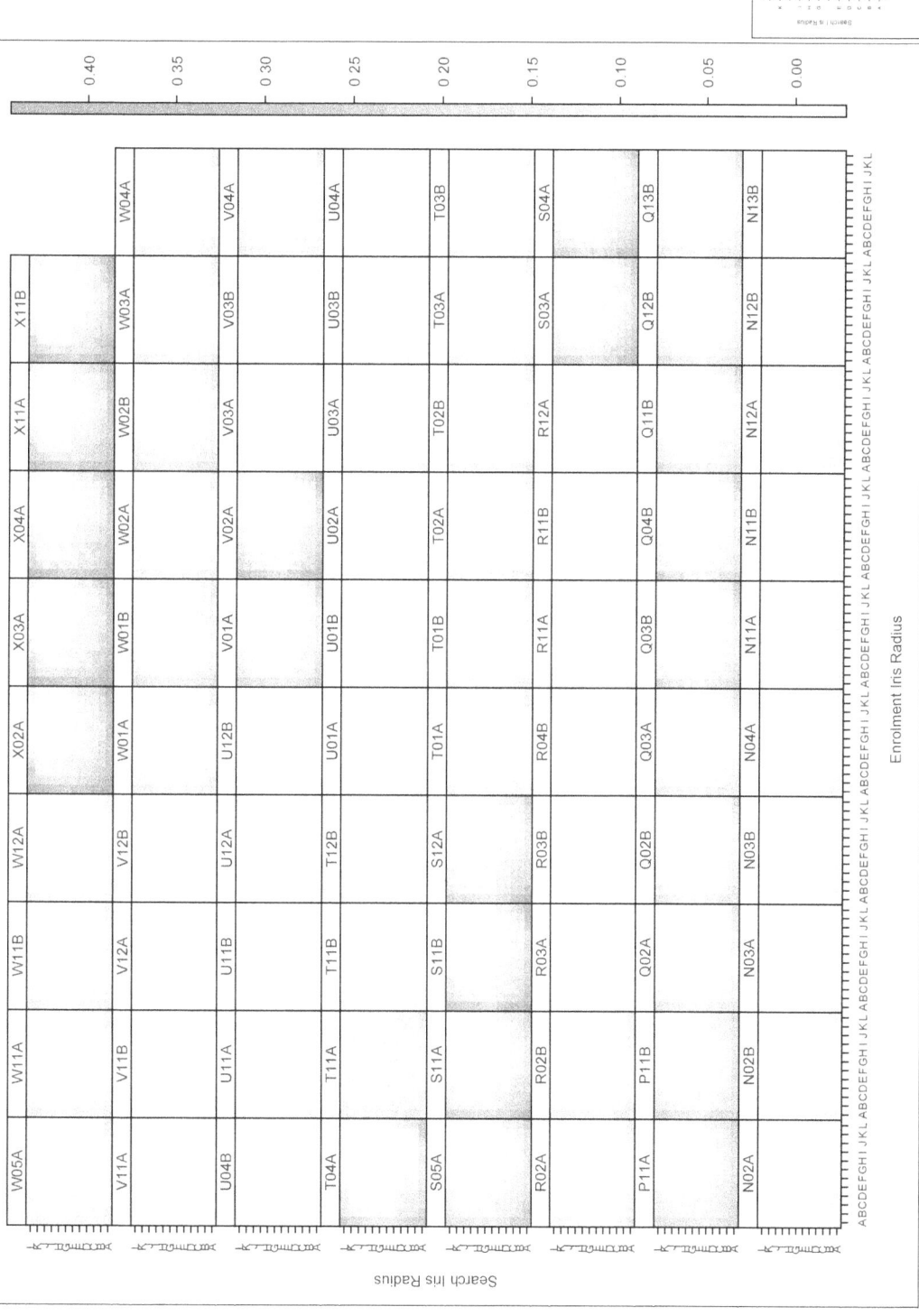

Figure 44: **Effect of radius:** Each panel shows FNIR for binned values of the iris radii of the enrollment and search images The radius bins A–L are listed in section 8.4.1 For this analysis, searches that did not produce the correct candidate were assigned a mate score equal to the highest observed mate score All mate scores were then jittered via additive rank-preserving Normal noise to break ties Normalization was the last step The mate searches were those of the large search set S_{1b} The enrolled population size was $N = 1,600,000$.

| FNIR = FALSE NEGATIVE IDENT. RATE | N = NEUROTECHNOLOGY | P = SMU | Q = IRITECH | R = COGENT | S = SMARTSENSORS | T = CAMBRIDGE |
| FPIR = FALSE POSITIVE IDENT. RATE | U = L1 | V = MORPHO | W = IRISID | X = CROSSMATCH | Y = KYNEN | |

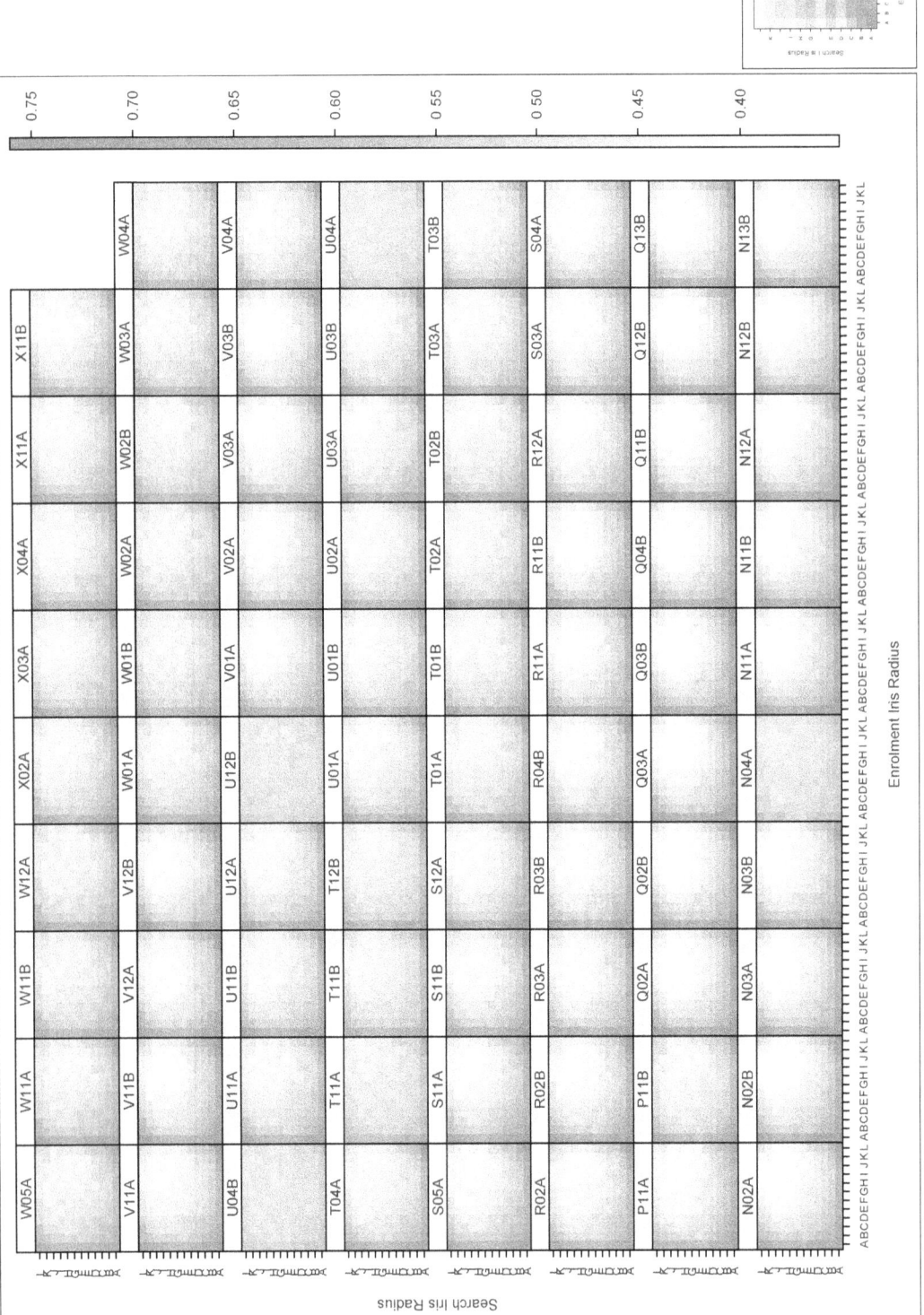

Figure 45: **Effect of radius:** Each panel shows median normalized mate dissimilarities for binned values of the iris radii of the enrollment and search images The radius bins A-L are listed in section 8 4 1 The normalized value for mate dissimilarity, x, is $0 \leq F(x) \leq 1$ where $F(x)$ is the cumulative distribution function for the mate scores for that algorithm This is done so that all panels color scales are on the same range For this analysis, searches that did not produce the correct candidate were assigned a mate score equal to the highest observed mate score All mate scores were then jittered via additive rank-preserving Normal noise to break ties Normalization was the last step The mate searches were those of the large search set S_{lb} The enrolled population size was $N = 1,600,000$

8.5 Quality

The ability to automatically inspect an image and produce a numerical estimate of the utility of the image to a downstream recognizer is an operationally desirable function. Particularly, if a low quality value is predictive of recognition failure (primarily a false negative, but possibly a false positive too), then a new sample can be collected while the subject is still present. This field of research is biometric quality assessment[25]. The quantitative relationship of quality values and specific measures derived from iris images has been addressed at length in the IREX II / IQCE report[30]. Here we conduct a far more limited study intended to determine whether the scalar quality values emitted during the course of normal image-to-template processing are related to false negative outcomes in one-to-many identification searches.

The quality values from the IREX III implementations are proprietary. That is, the image-specific measurements used in their computation are unknown.

8.5.1 Method

When N enrollment images are converted to templates, N quality values are computed. Likewise, during identification, image quality is assessed by the template generator ahead of one-to-many search. Some algorithms did not report quality values. Given M mated searches, the candidate lists will usually include a mate dissimilarity score d_i. In the cases where the mate is not found, we assign a poor mate score equal to the maximum value from all searches where the mate is found. Given search image quality values $Q_{si}, i = 1 \ldots M$, and their respective enrolled mate quality values Q_{ei}, we compute a combined image quality value as the geometric mean

$$Q_i = \sqrt{Q_{si} \, Q_{ei}} \tag{17}$$

This is an operationally irrelevant statistic because both images are not available when quality values are most wanted and useful i.e. during initial capture. The value is computed here because we seek to test intrinsic predictive power of the quality assessment algorithm. Here a restricted analysis (versus those of [30, 14]) quantifies the co-occurrence of a low quality value (from eq. 17) and a high (poor) dissimilarity mate score. Specifically we compute

$$\rho(r) = \frac{rM}{\sum_{i=1}^{M} \left(1 - H(Q_i - \tau_Q)\right) \, H(d_i - \tau_d)} \tag{18}$$

where r is the proportion of the lowest values we might not accept for further processing, $H()$ is the step function of eq. 2, τ_Q is a quality threshold taken from the cumulative distribution function of Q to select the lowest values $\tau_Q = F_Q^{-1}(r)$, and likewise τ_d is the dissimilarity above which there are rM mate scores via $\tau_d = F_d^{-1}(1 - r)$. Equation 18 quantifies the joint occurence of poor quality and poor mate scores. It is defined independently of operating threshold τ.

8.5.2 Results

Table 16 gives $\rho(r)$ values for 0.01×2^n for $0 \leq n \leq 5$. For $r = 0.01$ the best result, for algorithm Q02B, is that about one in seven misses were assigned low quality values ($\rho(0.01) = 0.142$). The ρ values increase with r such that for $r = 0.04$ the best result, from algorithm S12A, is that 39% of misses have low quality values.

8.6 Alternative quality combinations

Without reporting numbers, the following alternatives to eq. 17 generally give lower values for ρ than in the results section above.

---

[25] See the proceedings of two NIST quality workshops linked on the right side http://www.nist.gov/itl/iad/ig/ibpc2012.cfm.

▷ **Case 1**: $Q = Q_s$ - this is the case where an image is analyzed in isolation, or matched against other images in a sequence, and is operationally relevant because Q_e is assumed to be high because an image quality standard has been adhered to in all prior enrollments,

▷ **Case 2**: $Q = min(Q_e, Q_s)$ - the case where accuracy is driven by the lower of the two quality values; it is the same as above except that because the test data does not control enrollment quality at all, it gets closer to Case 1.

▷ **Case 3**: $Q = 1 - \sqrt{(1 - Q_e)(1 - Q_s)}$ - an ad hoc variant.

(a) Group 0

SDK	REJECTION FRACTION, r.					
	0.01	0.02	0.04	0.08	0.16	0.32
Q02A	0.148	0.181	0.191	0.225	0.316	0.465
Q02B	0.147	0.185	0.192	0.225	0.317	0.465
Q03A	0.117	0.144	0.174	0.255	0.357	0.492
Q03B	0.104	0.131	0.166	0.242	0.347	0.487
Q04B	0.111	0.137	0.166	0.242	0.347	0.487
R02A	0.026	0.040	0.081	0.146	0.233	0.390
R02B	0.030	0.050	0.084	0.141	0.231	0.381
R03A	0.026	0.045	0.090	0.149	0.233	0.388
R03B	0.031	0.046	0.074	0.128	0.221	0.375
R04B	0.050	0.079	0.097	0.143	0.229	0.378
S01B	-	-	-	-	-	-
S02B	-	-	-	-	-	-
S03A	0.014	0.024	0.038	0.067	0.124	0.285
S04A	0.109	0.217	0.291	0.319	0.246	0.325
S05A	0.000	0.000	0.303	0.324	0.246	0.325
T01A	0.039	0.070	0.112	0.193	0.329	0.531
T01B	0.078	0.112	0.156	0.222	0.351	0.543
T02A	0.034	0.060	0.112	0.217	0.366	0.553
T02B	0.062	0.095	0.149	0.242	0.371	0.552
T03A	0.079	0.112	0.163	0.252	0.380	0.556
T03B	0.072	0.107	0.162	0.248	0.373	0.536
T04A	0.032	0.064	0.108	0.188	0.336	0.544
U01A	0.085	0.115	0.195	0.298	0.445	0.619
U01B	0.081	0.121	0.197	0.299	0.446	0.620
U02A	0.070	0.115	0.195	0.298	0.445	0.619
U03A	0.013	0.026	0.048	0.095	0.208	0.402
U03B	0.016	0.026	0.047	0.099	0.213	0.410
U04A	0.016	0.026	0.051	0.102	0.192	0.343
U04B	0.018	0.027	0.052	0.101	0.187	0.341
V01A	0.031	0.059	0.093	0.149	0.226	0.363
V02A	0.020	0.031	0.064	0.125	0.205	0.353
V03A	0.073	0.090	0.134	0.209	0.311	0.442
V03B	0.070	0.086	0.127	0.203	0.303	0.435
V04A	0.038	0.058	0.100	0.173	0.281	0.420
W01A	0.047	0.068	0.118	0.194	0.307	0.500
W01B	0.019	0.028	0.060	0.114	0.232	0.476
W02A	0.057	0.074	0.119	0.189	0.301	0.496
W02B	0.021	0.029	0.061	0.115	0.232	0.476
W03A	0.050	0.069	0.119	0.189	0.302	0.496
W04A	0.046	0.058	0.098	0.171	0.292	0.492
W05A	0.049	0.059	0.101	0.171	0.292	0.492

(b) Group 1

SDK	REJECTION FRACTION, r.					
	0.01	0.02	0.04	0.08	0.16	0.32
Q11B	0.031	0.047	0.081	0.168	0.236	0.381
Q12B	0.030	0.051	0.081	0.168	0.236	0.381
Q13B	0.067	0.132	0.148	0.193	0.249	0.387
R11A	0.022	0.041	0.079	0.142	0.229	0.387
R11B	0.029	0.053	0.080	0.134	0.222	0.368
R12A	0.031	0.043	0.090	0.149	0.233	0.388
S11A	0.106	0.201	0.323	0.347	0.273	0.327
S11B	0.065	0.077	0.086	0.082	0.124	0.264
S12A	0.129	0.252	0.382	0.392	0.276	0.329
T11A	0.025	0.033	0.051	0.101	0.209	0.504
T11B	0.127	0.167	0.205	0.260	0.371	0.549
T12B	0.083	0.113	0.163	0.240	0.366	0.549
U11A	0.060	0.086	0.164	0.260	0.406	0.587
U11B	0.000	0.027	0.043	0.093	0.198	0.393
U12A	0.000	0.026	0.041	0.092	0.197	0.393
U12B	0.000	0.028	0.045	0.084	0.157	0.314
V11A	0.044	0.073	0.133	0.221	0.337	0.454
V11B	0.029	0.055	0.106	0.196	0.300	0.435
V12A	0.039	0.068	0.122	0.206	0.320	0.442
V12B	0.017	0.026	0.049	0.094	0.196	0.370
W11A	0.060	0.077	0.125	0.193	0.306	0.501
W11B	0.043	0.060	0.101	0.173	0.293	0.494
W12A	0.063	0.077	0.132	0.198	0.314	0.509

Table 16: Quality based rejection: The entries give the proportion of missed searches for which quality is low per equation 18. This is computed over the M mate searches in the large search set $\mathcal{S}_{1b}$, and an enrolled population size $N = 1,600,000$. Missing entries mean that quality assessment is not supported by the implementation, or that the run is incomplete. Cells are shaded dark and light green when the tabulated value is more than 10 and 5 times r, and red when the tabulated value is less than r. For this analysis, searches that did not produce the correct candidate were assigned a mate score equal to the highest observed mate score. Further, all mate scores were then jittered via additive rank-preserving Normal noise to break ties.

FNIR = FALSE NEGATIVE IDENT. RATE	N = NEUROTECHNOLOGY	P = SMU	Q = IRITECH	R = COGENT	S = SMARTSENSORS	T = CAMBRIDGE
FPIR = FALSE POSITIVE IDENT. RATE	U = L1	V = MORPHO	W = IRISID	X = CROSSMATCH	Y = KYNEN	

References

[1] R. M. Bolle, J. H. Connell, S. Pankanti, N. K. Ratha, and A. Senior. *Guide to Biometrics*. Springer, 2004.

[2] M Brauckmann and Busch C. *Large Scale Database Search*, pages 639–654. Springer-Verlag New York, Inc., Secaucus, NJ, USA, 2011.

[3] S. Curry, D. Founds, J. Marques, N. Orlans (Mitre), and C. Watson (NIST). Meds - multiple encounter deceased subject face database - nist special database 32. NIST Interagency Report 7679, National Institute of Standards and Technology, 2011. http://www.nist.gov/itl/iad/ig/sd32.cfm.

[4] J. G. Daugman. High confidence visual recognition of persons by a test of statistical independence. *IEEE Trans. Pattern Anal. Mach. Intell.*, 15(11):1148–1161, November 1993.

[5] John Daugman. Statistical richness of visual phase information: Update on recognizing persons by iris patterns. *International Journal of Computer Vision*, 45(1):25–38, October 2001.

[6] John Daugman. How iris recognition works. *IEEE Transactions on Circuits and Systems for Video Technology*, 14:21–30, 2002.

[7] John Daugman and Cathryn Downing. Epigenetic randomness, complexity and singularity of human iris patterns. *Proceedings of the Royal Society: B. Biological Sciences*, 268(1):1737–1740, 2001.

[8] L Dhir, N E Habib, D M Monro, and S Rakshit. Effect of cataract surgery and pupil dilation on iris pattern recognition for personal authentication. *Eye*, 24(6):1006–1010, 11 2009. http://www.nature.com/eye/journal/v24/n6/full/eye2009275a.html.

[9] J. Egan. *Signal Detection Theory and Analysis*. Academic Press, 1975.

[10] Jerome H. Friedman, Jon Louis Bentley, and Raphael Ari Finkel. An algorithm for finding best matches in logarithmic expected time. *ACM Trans. Math. Softw.*, 3:209–226, September 1977.

[11] P. Grother, G. W. Quinn, and P. J. Phillips. Evaluation of 2d still-image face recognition algorithms. NIST Interagency Report 7709, National Institute of Standards and Technology, 2010. http://face.nist.gov/mbe.

[12] P. Grother, E. Tabassi, G. W. Quinn, and W. Salamon. Irex i performance of iris recognition algorithms on standard images. Technical Report NIST Interagency Report 7629, National Institute of Standards and Technology, http://iris.nist.gov/irex/, October 2009.

[13] P. J. Grother and P. J. Phillips. Models of large population recognition performance. In *International Conference on Computer Vision and Pattern Recognition (CVPR)*, 6 2004. Contact: patrick.grother@nist.gov.

[14] Patrick Grother and Elham Tabassi. Performance of biometric quality measures. *IEEE Trans. Pattern Anal. Mach. Intell.*, 29(4):531–543, 2007.

[15] A. Gyaourova and A. Ross. A coding scheme for indexing multimodal biometric databases. In *Proc. of IEEE Computer Society Workshop on Biometrics at the Computer Vision and Pattern Recogniton (CVPR)*, Miami, FL, 6 2009.

[16] Feng Hao, J. Daugman, and P. Zielinski. A fast search algorithm for a large fuzzy database. *Information Forensics and Security, IEEE Transactions on*, 3(2):203 –212, june 2008.

[17] Zhaofeng He, Tieniu Tan, Zhenan Sun, and Xianchao Qiu. Toward accurate and fast iris segmentation for iris biometrics. *IEEE Trans. Pattern Anal. Mach. Intell.*, 31(9):1670–1684, September 2009.

[18] Mark D. Hill and Michael R. Marty. Amdahl's law in the multicore era. *IEEE Computer*, 7 2008.

[19] Karen Hollingsworth, Kevin W. Bowyer, and Patrick J. Flynn. Pupil dilation degrades iris biometric performance. *Computer Vision and Image Understanding*, 113(1):150–157, January 2009.

[20] J.P. Hube. Using biometric verification to estimate identification performance. In *Biometric Consortium Conference, 2006 Biometrics Symposium: Special Session on Research at the*, pages 1–6, 19 2006-aug. 21 2006.

[21] Daugman J. Probing the uniqueness and randomness of iriscodes: Results from 200 billion iris pair comparisons. *Proc. of the IEEE*, 94(11):1927–1935, 2006.

[22] Daugman J. New methods in iris recognition. *IEEE Trans. Systems, Man, Cybernetics B*, 37(5):1167–1175, 2007.

[23] J. Kittler, M. Hatef, R. Duin, and J. Matas. On combining classifiers. *IEEE Trans. Pattern Analysis and Machine Intelligence*, 20(3), March 1998.

[24] James R. Matey and Lauren R. Kennell. *Iris Recognition - Beyond One Meter*, pages 23–59. Springer Verlag, 1 edition, 2009.

[25] Donald M. Monro, Soumyadip Rakshit, and Dexin Zhang. Dct-based iris recognition. *IEEE Trans. Pattern Anal. Mach. Intell.*, 29:586–595, April 2007.

[26] Estefan Ortiz and Kevin W. Bowyer. Dilation aware multi-image enrollment for iris biometrics. In *International Joint Conference on Biometrics*, IJCB'11, Los Alamitos, CA, USA, 2011. IEEE Computer Society.

[27] P. Radu, K. Sirlantzis, G. Howells, S. Hoque, and F.Deravi. On combining information from both eyes to cope with motion blur in iris recognition. In *Proceedings of the Fourth International Workshop on Soft Computing Applications (SOFA)*, pages 175–181, July 2010.

[28] Roberto Roizenblatt, Paulo Schorr, Fabio Dante, Jaime Roizenblatt, and Rubens Belfort. Iris recognition as a biometric method after cataract surgery. *BioMedical Engineering Online*, 3(1), January 2004.

[29] Arun A. Ross, Karthik Nandakumar, and Anil K. Jain. *Handbook of Multibiometrics*. Springer, 2006.

[30] E. Tabassi, P. Grother, and W. Salamon. Irex ii : Iqce - performance evaluation of iris quality measures. Technical Report NIST Interagency Report 7820, National Institute of Standards and Technology, http://iris.nist.gov/irex/, September 2011.

The IREX III Appendices

In addition to appendices A and B that appear in this document, the IREX III report is supplemented with appendices C through I, which form the IREX III APPENDICES. They are provided as a single PDF file extending to several hundred pages. It contains exhaustive algorithm-specific results that have been generated by an iterating script. It will be of primary interest to the algorithm developers, and to users considering particular algorithms.

The PDF and a compressed ZIP are available from the IREX homepage http://iris.nist.gov/irex.

▷ **Appendix A**: Comparison of Iris and Face Recognition Performance

▷ **Appendix B**: Background on Accuracy Metrics

▷ **Appendix C**: Threshold Calibration Plots

▷ **Appendix D**: Effect of Population Size on DET

▷ **Appendix E**: Effect of Compression

▷ **Appendix F**: Effect of Pupil Dilation and Contriction

▷ **Appendix G**: Relation of Image Quality Estimates and Accuracy

▷ **Appendix H**: Effect of Image Size and Camera

▷ **Appendix I**: Relationship between Score Distributions and Search Duration

▷ **Appendix J**: Implementation Traceability

A Comparison of Iris and Face Recognition Performance

A.1 Introduction

In 2010, NIST ran the MBE-STILL evaluation of one-to-many identification algorithms[11]. Over a nine month period, the activity measured performance of approximately twenty of the core enrollment and search algorithms from the leading and emerging commercial providers of face recognition technology. The test used two sequestered operationally-collected datasets; one (DOS) from visa applications, and another from persons detained as part of routine law-enforcement (LEO) operations. As detailed below, the use of the latter affords an excellent opportunity to compare face and iris performance because the images have excellent ground-truth integrity stemming from their collection-time pairing with ten fingerprint records.

A.2 Details of the comparison

The experimental method is as follows. For face exactly 1.6 million images of 1.6 million individuals were enrolled; these were drawn randomly from the LEO-A mugshot population[26]. Examples of such images are shown in Figure 48 and are available from NIST as Special Database 32[1]. For iris exactly 1.6 million images of 1.6 million individuals were enrolled, as described in section 3.1 of this report. Image and subject-specific metadata information (such as sex, age, left-right eye label) was not provided to the algorithms under test.

For face, the number of mated searches was 25,000 involving exactly 25,000 individuals. The number of nonmate searches was 200,000 using 200,000 individuals. For iris, the number of mated searches was 238,740 involving 71,351 individuals, some people and eyes being imaged on several occasions. The number of nonmate searches was 311,427. This involved 160,000 persons, using no more than one image per eye. The number of searches influences the size of the confidence interval associated with random error. The size of the enrolled population impacts accuracy itself, primarily the false positive rate, but also rank-based measures.

A.3 Results

Accuracy: Figure 46 shows accuracy of face and iris identification systems as a detection error tradeoff characteristic DET. For identical enrolled population sizes of N = 1.6 million, this plots, on logarithmic axes, the false negative identification rate (FNIR) against false positive identification rate (FPIR) which are computed, for any given threshold, as fractions of mates and nonmates that are not reported and reported, respectively, on candidate lists.

The notable observations are as follows

 ▷ The iris recognition algorithms give much lower error rates than those for face. For any fixed target FPIR, the best face and iris algorithms give FNIR rates an order of magnitude apart. For example, at FPIR= 0.0001 face algorithms V21 and W22 miss about 20% of mates while iris algorithms U12B and V11B miss less than 2.5%.

 ▷ This difference is larger at the lower FPIR values needed for automated identification. At high FPIR, when false positives are tolerated, for example when a human examiner will adjudicate items on candidate lists, face recognition accuracy is closer to that of iris. Thus, at FPIR= 1, the best face implementation, V21, offers a miss rate of 5% just better than the least accurate iris algorithm included in the graphs. The best iris algorithm misses about 1.3%. These rates critically depend on the fraction of non-ideal images in the datasets, and on the correctness of the ground-truth identities, see 6.3.

---

[26]The law enforcement (LEO) photos are comprised of two populations, LEO-A and LEO-B. The LEO-A population was collected with variable conformance to documentary mugshot standards. The LEO-B photos were collected using web-cameras; they exhibit distortion artifacts, and are far from conformance to illumination and other standardized quality aspects. The LEO-B images are excluded from this analysis because they are clearly and systematically not representative of what a day-forward face recognition application would aim to use.

| FNIR = FALSE NEGATIVE IDENT. RATE | N = NEUROTECHNOLOGY | P = SMU | Q = IRITECH | R = COGENT | S = SMARTSENSORS | T = CAMBRIDGE |
| FPIR = FALSE POSITIVE IDENT. RATE | U = LI | V = MORPHO | W = IRISID | X = CROSSMATCH | Y = KYNEN | |

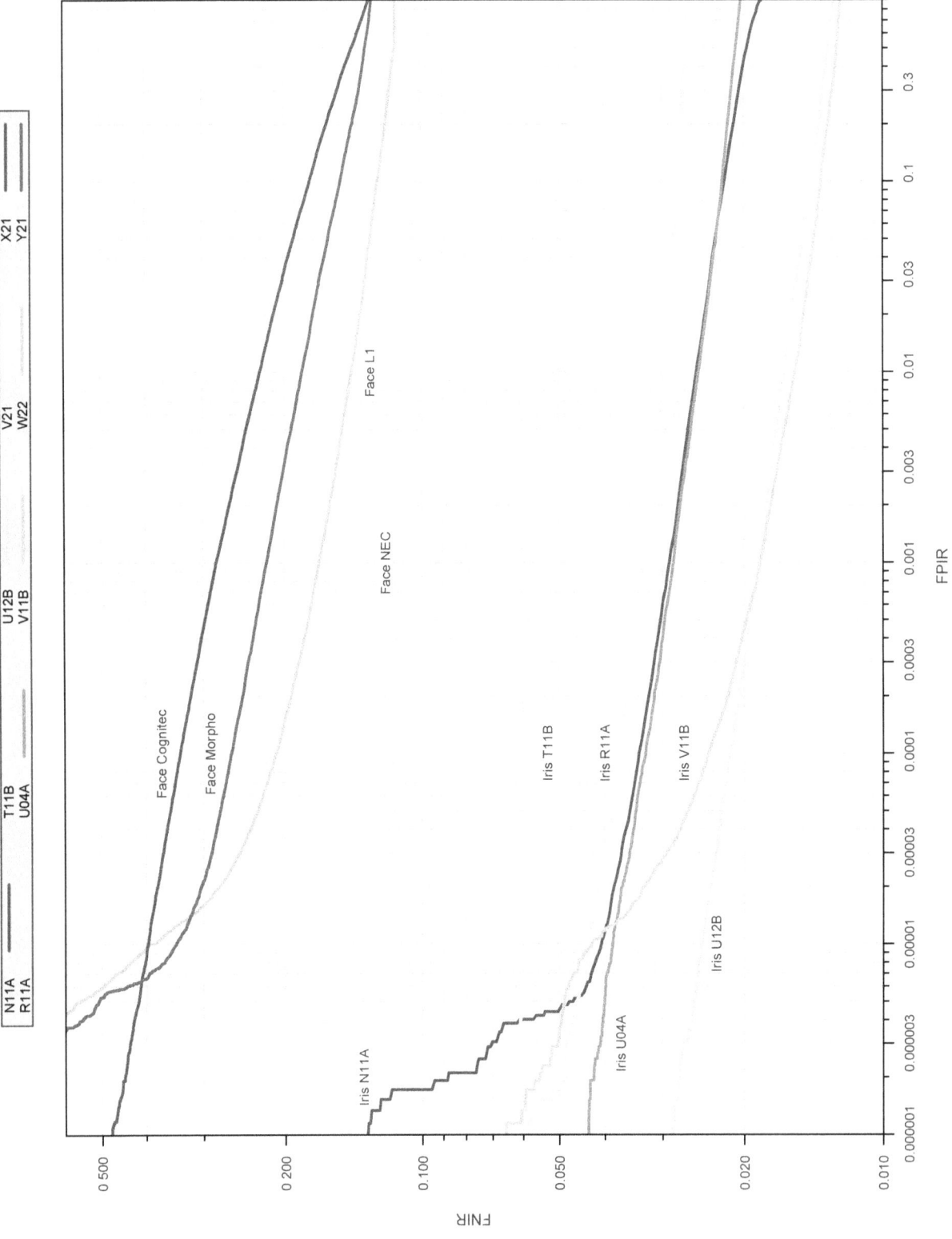

Figure 46: Accuracy of face (above, green and blue colors) and iris (below, yellow, orange, red and brown colors) recognition algorithms   The face algorithms are identified by three characters, the first of which identifies the provider: X = Cognitec, Y = Morpho, V = NEC, and W = L1   The iris algorithms are identified by four characters as in the rest of this report; providers are identified in the running footer   The implementations execute one-to-many searches in populations of size N = 1,600,000 persons, one face per person, two irides per person but enrolled as though they were from different persons (i e   under different identifiers)   The plot is a 1:N DET - FPIR is logically N times an implied 1:1 FMR such that the left side of the DET corresponds to false match rates below 10^{-11}

| FNIR = FALSE NEGATIVE IDENT. RATE | N = NEUROTECHNOLOGY | P = SMU | Q = IRITECH | R = COGENT | S = SMARTSENSORS | T = CAMBRIDGE |
| FPIR = FALSE POSITIVE IDENT. RATE | U = L1 | V = MORPHO | W = IRISID | X = CROSSMATCH | Y = KYNEN | |

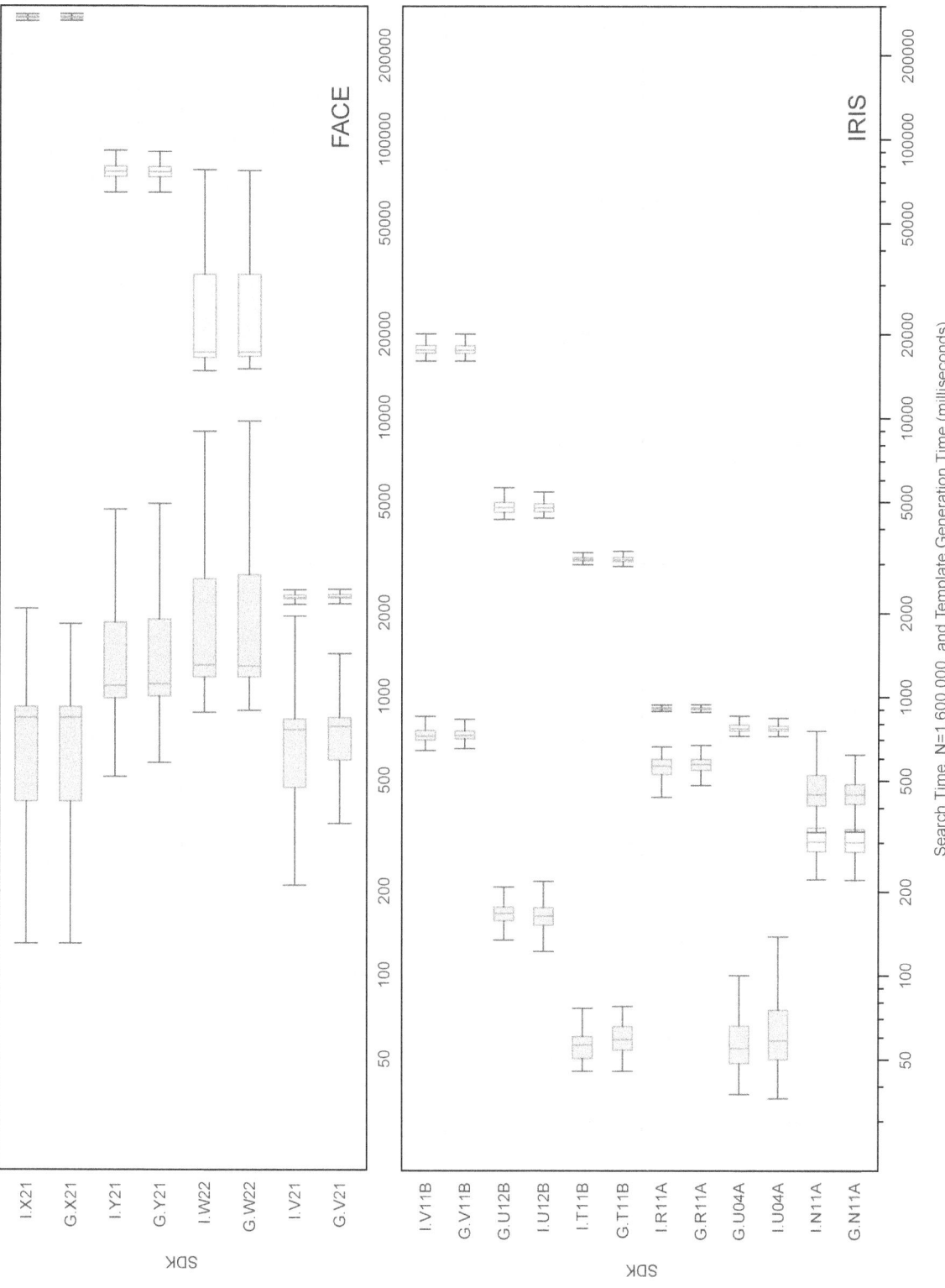

Figure 47: Comparison of the computational expense of face (above) and iris (below) recognition algorithms The green boxes (typically on the left) show template generation duration The gold boxes (right) show one-to-many search durations for N = 1,600,000 persons, one face per person, two irides per person but enrolled as though they were from different persons (i e under different identifiers) The algorithm identifier at left is accompanied by a "G", indicating genuine mated search, or an "I" indicating impostor nonmate search - this only makes a difference for T12B The providers of face algorithms are coded as X = Cognitec, Y = Morpho, V = NEC, and W = L1

| FNIR = FALSE NEGATIVE IDENT. RATE | N = NEUROTECHNOLOGY | P = SMU | Q = IRITECH | R = COGENT | S = SMARTSENSORS | T = CAMBRIDGE |
| FPIR = FALSE POSITIVE IDENT. RATE | U = L1 | V = MORPHO | W = IRISID | X = CROSSMATCH | Y = KYNEN | |

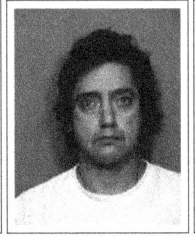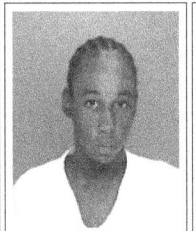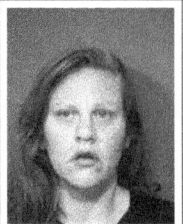

Figure 48:   Examples of face images from three subjects. While geometry placement is good, variations in pose and expression represent mild deviations from the appearance requirements of formal standards such as ISO/IEC 19794-5:2005 or 2011. Images are of size at least 480x600 rising to 720x960 pixels. The reader should understand that appearance will depend on display and printer properties.

▷ These figures can be restated for the case of fixing FNIR at, say, 5% i.e. a hit rate of 95%. If such an operating point were necessary, the FPIR of the leading iris algorithms is below 10^{-5} and that of the one face algorithm capable of find this many mates is close to 1. That means that five orders of magntitude (i.e. factor larger than 100,000) more nonmates will be generated with single face versus single iris. Practically, face identification systems are incapable of accessing the same accuracies as iris *given the respective image qualities*. Note that if the face collection practices were better, with better conformance to the subject appearance requirements of standards, particularly regarding frontal pose, the face DET curves would improve. Iris accuracy would likewise improve given tighter acquisition controls.

▷ Note that the gradient of the DET curves is sometimes cited as a statement of the discriminative power of a biometric, because it is equal to the likelihood ratio[9]. While a naïve inspection of Figure 46 reveals the trace of the X21 face algorithm to have similar slope to that of the iris algorithms, this is incorrect. The use of log-log scales means that the gradient of the U12B iris DET is actually about a factor of 20 lower than that of the X21 face algorithm.

Speed: Regarding the computational expense plots of Figure 47 the following observations are evident.

▷ Iris template generation is typically five to ten times faster than that for face. A small part of this is attributable to the size of the image itself - fewer pixels are processed faster.

▷ Iris search is typically faster than face search: the fastest iris search algorithm (T12B) is ten times quicker than the fastest face algorithm (V21). Also, while there is factor of ten disparity in speed between the slowest iris and face algorithms (V11B and X21), the most accurate face algorithm (V21) is faster than the most accurate iris algorithm (V11B).

▷ Not reported here is the dependence of speed on N. If we use a power-law aN^b to model this, then most iris algorithms exhibit a linear, $b = 1$, dependence (see Figures 7 and 8, and also 12) while many face algorithms show better scalability $b < 1$ (see the MBE-STILL report[11]).

A.4   Discussion

The accuracy results show iris as superior to face; this result will be sustained only to the extent that the input data and algorithms are similar to those used here. Additionally, while core accuracy is influential on virtually all applications of face and iris technology, a large variety of other factors are relevant to operational viability and outcomes. Among these are technical and non-technical issues including:

▷ **Application and policy constraints**: Choice of modality may depend on external constraints such as legal, privacy, sociological policies or regulations, existing practice, interoperability requirements with other parties, or whether a human may need to review images.

▷ **Image size**: Transmission, and storage and retention requirements may influence whether iris or face is more appropriate. Face images on unpowered identity credentials ("smartcards") are typically specified to have sizes exceeding 15KB while iris image sizes down to 2KB have dedicated support in formal standards.

▷ **Template size**: Depending on the application, the size of the feature data from an image is primarily related to one-to-one and one-to-many recognition speed (for reasons of memory-to-CPU bus bandwidth and algorithmic complexity), on transmission cost (from collection point to backend system), and on storage cost (on smart cards or on disks).

▷ **Search duration**: The time needed to execute a search varies greatly depending on algorithm (see section 4.5). This is obviously a component of response time for searches.

▷ **Scalability**: The term covers several aspects related to the dependence of accuracy, speed and resource usage as the enrolled population grows. The first order assumption is that these quantities scale linearly with population size. For iris recognition, Table 10 shows that algorithmic cost increases about linearly with enrolled population size, N. For face recognition, however, the search speed of some algorithms grows better than linearly aN^b, with $b < 1$, such that a tenfold increase in N produces only a 10% increase in search time[27].

▷ **Availability**: The face is an overt biometrics: it can be collected at a distance, with zero or minimal cooperation of the subject. Iris also can be collected at a distance, though the resolution requirements alone necessitate more costly imaging apparatuses. All applications typically require, and always benefit from, a fully cooperative presentation. Thus, electronic passport face verification applications benefit from tight specifications on facial frontal pose and illumination. Iris images, too, are more likely to enroll and be recognized when the subject looks directly at the camera.

▷ **Ubiquity**: One of the fundamental properties for a biometric is that all members of the population possess or express the biometric. For iris, a small fraction of the population has no discernible iris - this condition, Aniridia, is estimated to affect no more than 40000 persons[28]. More prosaically, iris may be occluded by sunglasses or patterned contact lenses, while face is sometimes covered by a scarf or veil.

▷ **Disease**: A wide range of medical conditions affect the eye and some of these may impede iris recognition. Some conditions are very rare. As a larger anatomical structure, the face is probably less affected by abnormalities that would undermine detection and recognition. Research is underway on the effect of atypical irides ([ ], other studies are as yet unpublished). For both biometrics and any given non-ideality, the following issues are relevant to recognition:

  – **Nature** - Whether the condition is acute and temporary, or chronic and persistent;

  – **Prevalence** - How common is the condition;

  – **Impact** - Is recognition undermined and, if so, is it relevant in the application;

  – **Detection** - Can the condition be detected so as to initiate secondary action;

  – **Mitigation** - Can recognition be improved by appropriate re-enrollment;

▷ **Camera capability**: Recently face cameras have been equipped with face detection algorithms, and this can improve focus and exposure control. However, face images have historically been collected by non-active cameras, and quality control has been enforced only with human review. For iris, cameras have been designed almost universally to collect iris images specifically by providing a dedicated illuminant and some real-time iris detection. This is often effective at ensuring that the eye is focused, centered, not moving, and open. The time needed to execute an iris localization and segmentation algorithm is less than the time needed to fully generate a template. The times given in Table 8 would support iris localization at several frames per second on a camera's embedded processor. Iris localizaton would allow iris specific exposure measurements, for example. Full-blown template generation inside the camera would allow matching of sequential frames and output of the best matching image.

---

[27] See Figure 11 in NIST Interagency Report 7709 linked from http://face.nist.gov/mbe
[28] Curiously, despite the large population sizes, none were noted in the IREX III failure analysis

FNIR = FALSE NEGATIVE IDENT. RATE	N = NEUROTECHNOLOGY	P = SMU	Q = IRITECH	R = COGENT	S = SMARTSENSORS	T = CAMBRIDGE
FPIR = FALSE POSITIVE IDENT. RATE	U = LI	V = MORPHO	W = IRISID	X = CROSSMATCH	Y = KYNEN	

▷ **Cost**: Iris is commonly held to be more expensive than face recognition. In large part this reflects that iris cameras are purpose-built. In any case, sensor cost is but one part of the application cost. A complete accounting would include consideration of many elements including procurement cost, capital cost of front end camera and client equipment, capital cost of backend infrastructure (if any), recognition and compression algorithm licensing, integration costs, fixed and variable communications costs, maintenance costs, costs associate with recognition errors, operator cost in enrollment and recognition phases, integration costs for networks, cryptographic support, and forensic resolution of candidates and errors.

B   Biometric Error Rate Tradeoff Characteristics

This Appendix is intended to give a biometric identification-specific overview of the Detection Error Tradeoff characteristic DET. More general and detailed information is given in the Egan's class book[9].

Accuracy Terms + Definitions

A **detection error tradeoff** (DET) characteristic represents the tradeoff between Type II and Type I classification errors. A **receiver operating characteristic** (ROC) is usually equivalent and the terms are synonymous. In biometrics, Type II errors occur when two samples of one person do not match – this is called a **false negative**. Correspondingly, Type I errors occur when samples from two persons do match – this is called a **false positive**. Matches are declared by a biometric system when the native comparison score from the recognition algorithm meets some **threshold**. Comparison scores can be either **similarity scores**, in which case higher values indicate that the samples are more likely to come from the same person, or **dissimilarity scores**, in which case higher values indicate different people. Similarity scores are traditionally computed by **fingerprint** and **face** recognition algorithms, while dissimilarities are used in **iris recognition**. In some cases, the dissimilarity score is a distance; this applies only when **metric** properties are obeyed. In any case, scores can be either **mate** scores, coming from a comparison of one person's samples, or **nonmate** scores, coming from comparison of different persons' samples. The words **genuine** or **authentic** are synonyms for mate, and the word **impostor** is used a synonym for nonmate. The words mate and nonmate are traditionally used in identification applications (such as law enforcement search, or background checks) while genuine and impostor are used in verification applications (such as access control).

For iris recognition, mate comparisons yielding dissimilarities greater than a threshold are false negatives. In identification these are called **misses** and contribute to the **false negative identification rate** (FNIR). Nonmate comparisons at or below a threshold are false positives; in identification these are sometime called **false alarms**, and they contribute to **false positive identification rate** (FPIR). The threshold can take on any real value, and it is conventional in biometrics testing to examine error rates as a function of the threshold. In many systems, the threshold can be varied continuously, while in other (production) systems, it may only take on a few settings.

Returning to the DET, it plots a function of FNIR against a function of FPIR. Here and in many other reports, the function is the logarithm function (log axes). However, a DET might also plot the **hit rate**, and the true positive identification rate, TPIR = 1 − FNIR is plotted on a linear scale; this is often referred to as a ROC. More rarely, the function might be the inverse Gaussian function.

More detail and generality is provided in formal biometrics testing standards, see the various parts of ISO/IEC 19795 Biometrics Testing and Reporting. More terms, including and beyond those to do with accuracy, see ISO/IEC 2382-37 Information technology -- Vocabulary -- Part 37: Harmonized biometric vocabulary

---

FNIR = False Negative Identification Rate
FNIR = FNIR(N, T, L, R)
FNIR is computed by executing mate searches into an enrolled population of size N. It is the proportion of mate searches for which the mate is
- EITHER   not returned as any of L candidates,
- OR   is present but has dissimilarity above threshold T
- OR   is present at rank greater than R.

In IREX III, the rank criterion is not used for DET computations, i.e. $R \rightarrow \infty$, so FNIR is solely a function of population size, N and threshold, T. FNIR(N, T).

FPIR = False Positive Identification Rate
FPIR = FPIR(N, T, L)
FPIR is computed by executing nonmate searches into an enrolled population of size N. It is the proportion of returned candidates which have dissimilarity at or below threshold T. If S searches are conducted, S x L candidates will be returned, and FPIR is the number at or below threshold, divided by (S x L).

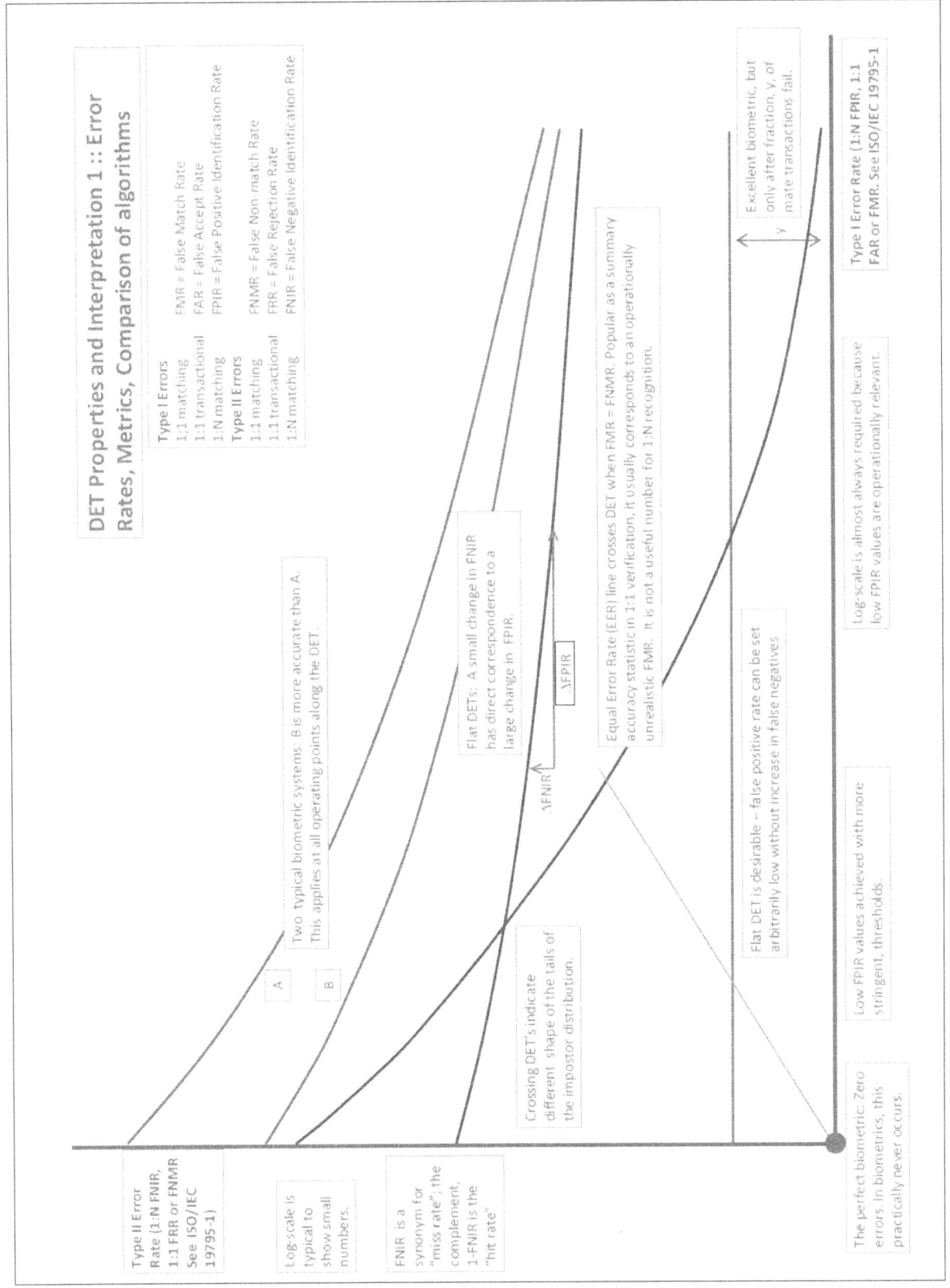

Figure 49:

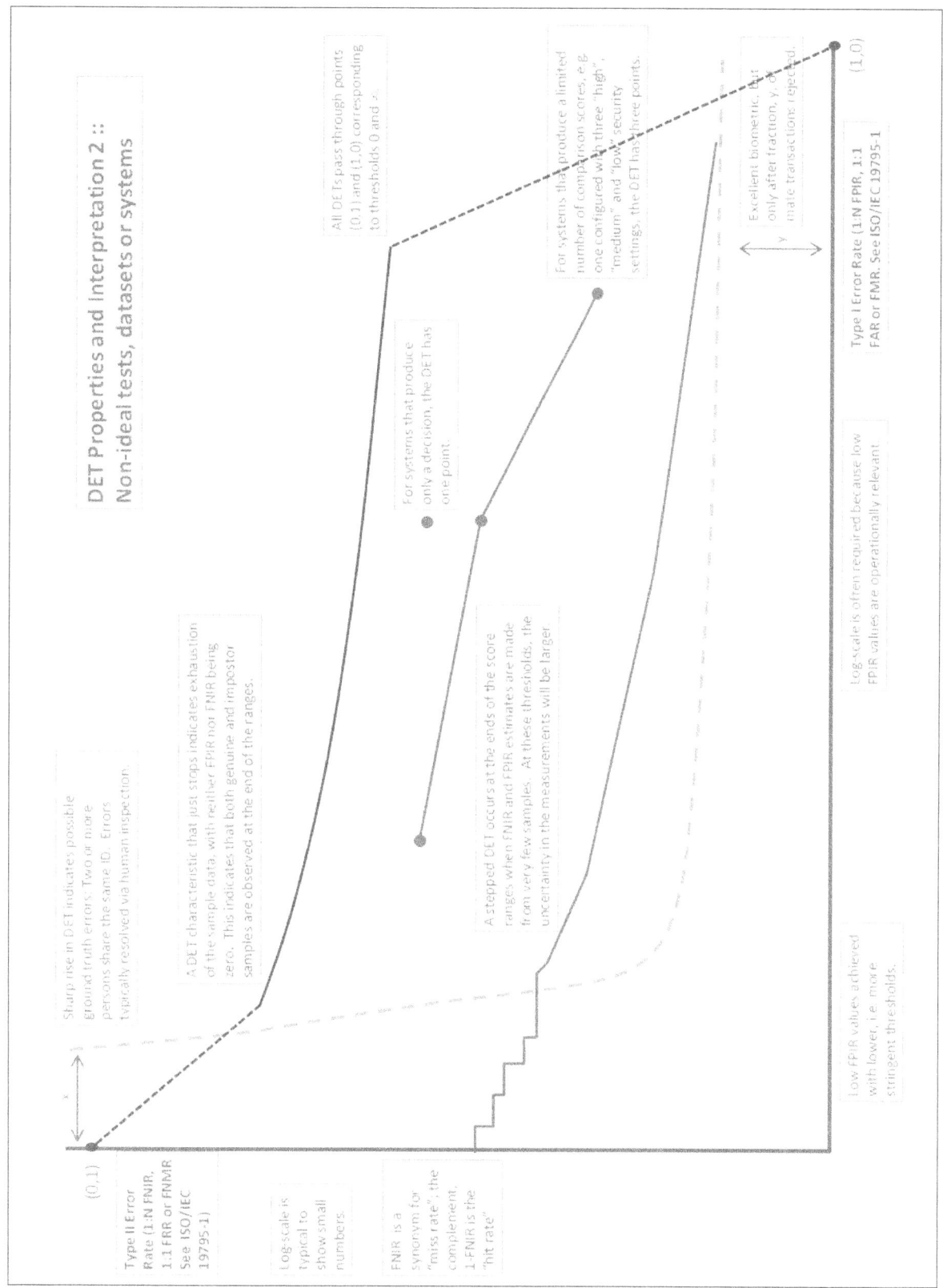

Figure 50:

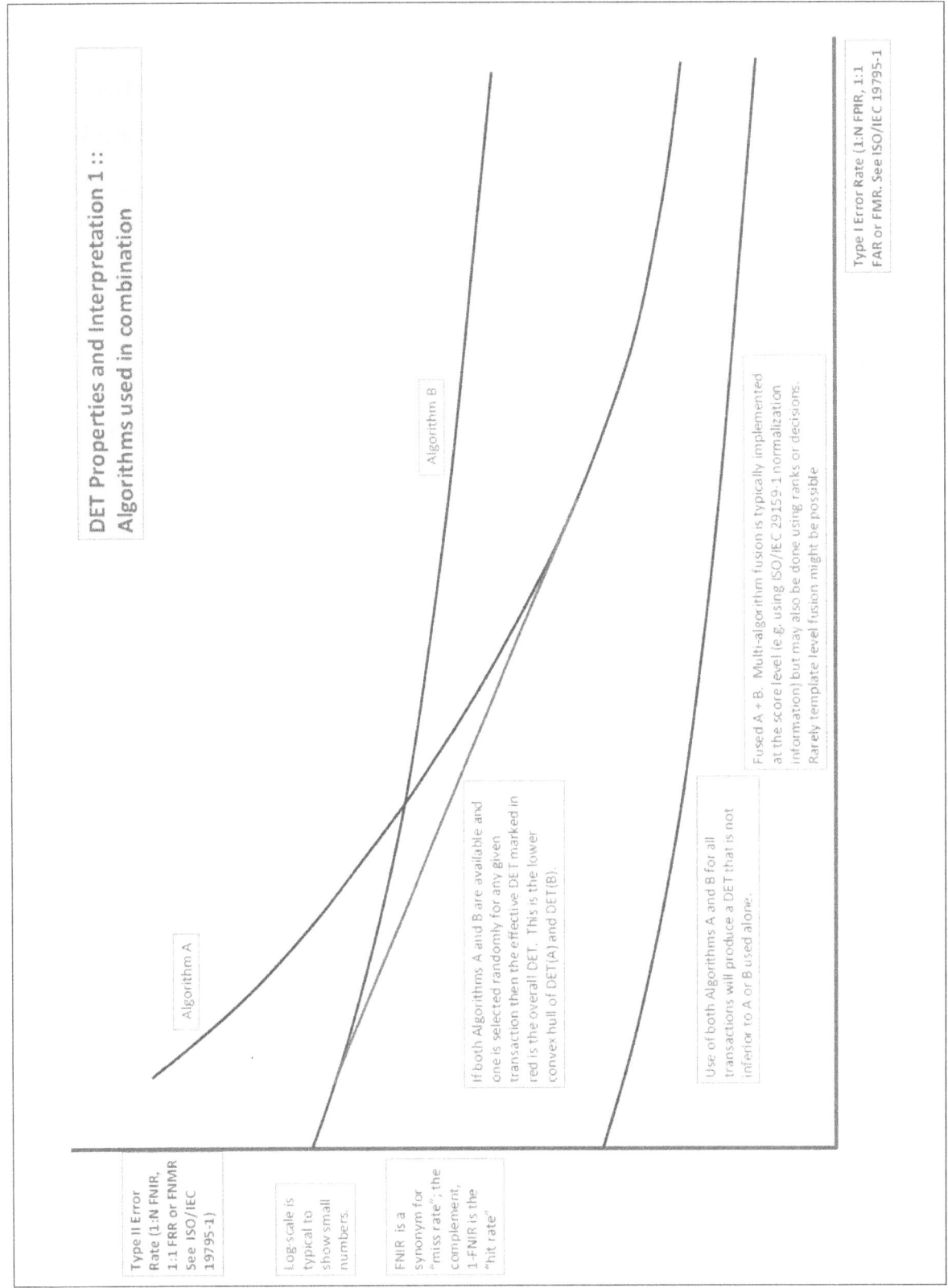

Figure 51: